高等院校土建学科双语教材（中英文对照）
◆ 城市规划专业 ◆
BASICS

城 市 分 析
URBAN ANALYSIS

[德] 格里特·施瓦尔巴赫　编著

杨　璐　柳美玉　　译

中国建筑工业出版社

著作权合同登记图字：01-2009-7699 号

图书在版编目（CIP）数据

城市分析/（德）施瓦尔巴赫编著；杨璐，柳美玉译. —北京：中国建筑工业出版社，2011. 11
高等院校土建学科双语教材（中英文对照）◈ 城市规划专业 ◈
ISBN 978-7-112-13647-6

I. ①城… II. ①施…②杨…③柳… III. ①城市规划-双语教学-高等学校-教材-汉、英 IV. ①TU984

中国版本图书馆 CIP 数据核字（2011）第 204842 号

Basics：Urban Analysis / Gerrit Schwalbach
Copyright ⓒ 2009 Birkhäuser Verlag AG（Verlag für Architektur），P. O. Box 133，4010 Basel，Switzerland
Chinese Translation Copyright ⓒ 2012 China Architecture & Building Press
All rights reserved.
本书经 Birkhäuser Verlag AG 出版社授权我社翻译出版

责任编辑：孙书妍
责任设计：陈　旭
责任校对：张　颖　王雪竹

高等院校土建学科双语教材（中英文对照）
◈ 城市规划专业 ◈
城市分析
[德] 格里特·施瓦尔巴赫　编著
杨　璐　柳美玉　译
＊
中国建筑工业出版社出版、发行（北京海淀三里河路9号）
各地新华书店、建筑书店经销
北京雅盈中佳图文设计公司制版
北京建筑工业印刷厂印刷
＊
开本：880×1230毫米　1/32　印张：4⅝　字数：190千字
2012 年 4 月第一版　　2018 年 10 月第二次印刷
定价：29.00 元
ISBN 978-7-112-13647-6
　　（32403）
版权所有　翻印必究
如有印装质量问题，可寄本社退换
（邮政编码　100037）

中文部分目录

CONTENTS

序

　　对现有城市结构及其所具有的不同社会、历史、建筑影响进行分析是现代城市规划的一项主要内容。城市以及城市区域的特征往往是在许多代人的影响下所形成的，而不同代人之间的想法各不相同。所以说，城市邻里的特征是在以上不同因素的共同作用下所形成的。

　　对现有城市邻里的特征本质进行分析，探明其更广范围的空间情境以及城市规划中的错误和失败之处，是确保城市邻里未来长期健康发展的基础工作。因此，城市分析对于城市规划而言至关重要，同时也是在实际工程实施过程中进行合理规划的基础。

　　本套书中关于城市规划的几册书将向城市规划和建筑学专业的学生介绍相关的基础知识以及实际工程中的不同方法。本书是高等院校土建学科双语教材中《城市街区》一册的补充，主要介绍城市分析的不同元素和对已有城市结构及其特征的分析。重点内容包括：进行不同城市元素分析的准备工作和方法、数据和相关信息的处理以及特定领域的分析工作。作者根据自身的实际工程经验，介绍了进行城市分析的方法，论证了分析结果类型，并根据分析结果提出了可能的应对措施。本书将理论方法和实践知识很好地结合在一起，同时具有很好的学术性和专业性。

编者：贝尔特·比勒费尔德（Bert Bielefeld）

FOREWORD

Studying existing urban structures, with their diverse social, historical and architectural influences, is a major part of modern urban planning. Cities and city districts are often marked by many generations, each with a different way of thinking. An urban neighborhood is characterized by the way these many influences overlap.

Analyzing the mechanisms behind an urban neighborhood's characteristics, recording its broader spatial context, and identifying the failings and mistakes of urban planning form the basis for safeguarding the long-term future of existing urban neighborhoods and their further development. This makes city analysis fundamental to any urban planning project. It is also the basis for intelligent urban planning context handling when implementing actual construction projects.

The Basics books on urban planning provide a basic grounding and various practical working methods for students of urban planning and architecture. Supplementing Basics Urban Building Blocks, an introduction to individual urban structural elements, the subject of this book is the analysis of existing urban structures and their characteristics. The emphasis is on explaining how to prepare and implement analyses of all urban factors, deal with sources of data and information and work in a specific area of analysis. The author presents methods of analysis from his own practical experience, demonstrates analysis result documentation types and reviews possible courses of action based on these results. Basics Urban Analysis is a valuable combination of methodical approaches and practical knowledge that students will find useful both academically and professionally.

Bert Bielefeld
Editor

INTRODUCTION

Urban analysis
and
urban planning

Urban planning always demands analysis of the area involved. This provides the background knowledge and rationale for any urban planning measure. Urban analysis, the study of existing urban spaces, is the preliminary stage of urban planning in any existing city.

Urban planning has increasingly less to do with quantitative growth and more to do with caring for and adapting existing urban structures, especially in Europe. This change in orientation requires intensive study of existing urban structures. The fundamental question, however, is whether existing cities' future development can actually be planned, and whether this kind of development can be determined in advance by analyzing the initial conditions. › Fig. 1

Urban planning has been determining future demands on urban spaces by measuring particular values for a long time. This approach

Fig.1:
Town planning within organic urban structures requires a thorough urban analysis.

9

intensified with the coming of the 20th century. During the Industrial Revolution, many cities experienced almost totally unplanned expansion. Experts noted their functional deficiencies. The planning approach developed as a response was particularly strongly represented in the 1960s, as the emergence of computers enabled extensive data processing and made advance planning of urban development with precision look like a realizable goal. The role of the urban planner also changed, with the unaccountable master builder becoming a rational engineer who would diagnose the city and then prescribe developmental planning measures. However, all efforts to precisely predetermine a city's development failed. Today, urban development processes are considered far too complex and contradictory to be understood and extrapolated in their entirety. Despite this realization, urban planning and urban analysis remain indispensable, but they are not used to plan every detail of urban development. Instead, they provide a basis for integrating diverse individual developments into a single planning scheme. With urban planning increasingly focusing on existing infrastructure, urban analysis has a key role. Before a need for action can be discovered and appropriate stabilization measures can be put in place, existing urban structures must be analyzed. Urban analysis does not precede urban planning in a fixed, self-contained process; rather, they are both elements in a continuous process. › Fig. 2

Urban analysis

Urban analysis describes or analyses particular factors relevant to urban planning, in whole cities or in smaller areas within cities. Aesthetic, spatial, social or economic factors are recorded and represented in the study, which goes on to describe the interactions between these effects – which, however, cannot be assessed or described in full because the city and its users interact in a reciprocal system. Urban analysis is therefore confined to abstract, model interpretations of the city or its individual areas. › Fig. 3

Cities are generally perceived in a very different way by different parties. Factors like the emotional ties of residents to their neighborhood or their place in the local society of a certain neighborhood are incomprehensible to outsiders. Conversely, faults in urban design etc. may be of only minor importance to local residents. The urban planning process regularly involves serious conflict between urban planners and residents or users of the city due to their different perceptions and evaluations of a city. Initial conflicts are unavoidable, but if handled properly they represent a chance to develop a deeper understanding of the situation within the space under observation, spurring residents to look beyond the context of their own neighborhood and giving urban planners an insight into that neighborhood's social mechanisms. › Fig. 4

Fig.2:
Many individual developments come together in an organic urban structure.

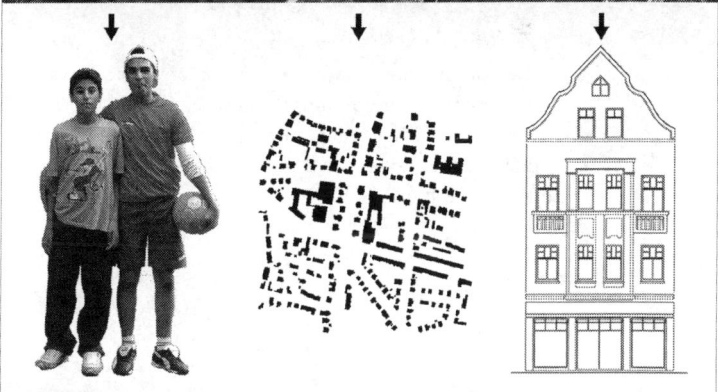

Fig.3:
Urban analyses do not reproduce a city in its entirety.

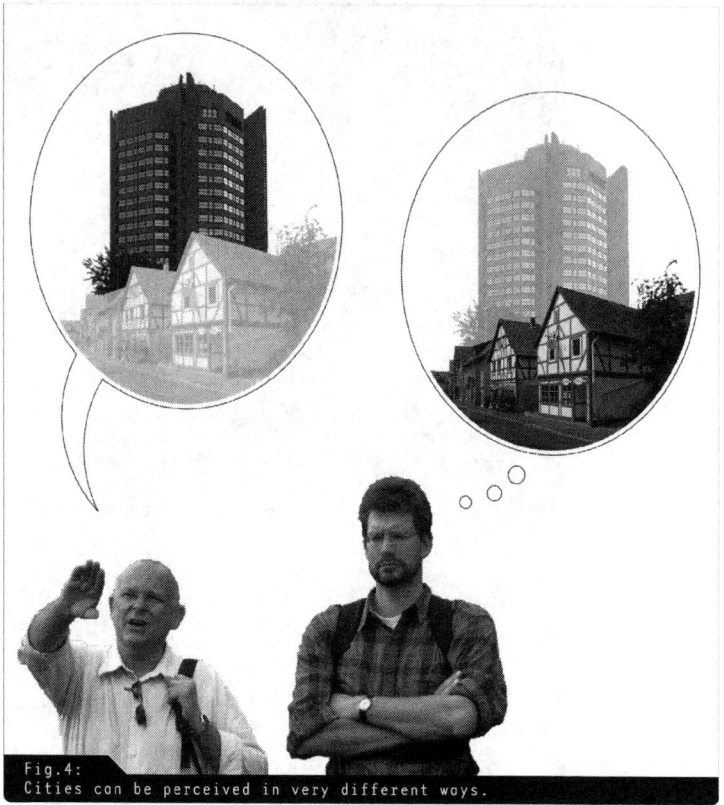

Fig.4:
Cities can be perceived in very different ways.

Urban analysis may be part of a formal planning process. In the context of public development programs or urban renovation or renewal, an urban analysis may be implemented according to specific requirements. In general, however, the implementation of an urban analysis is not standardized; i.e. its extent and degree of detail must be decided based on the individual situation.

A city can be represented as a tangible, perceptible spatial phenomenon, without precisely analyzing or describing its causative factors. Conversely, a city can be seen as the field of activity for scientifically measurable factors, without reference to its spatial dimension. Spatial analyses generally avoid both of these extreme approaches, instead describing a city as an interaction of different effects, but all with spatial implications. The main reference point is usually the perceptible city – urban analysis cannot exist without reference to tangible, perceptible phenomena.

AESTHETIC OBSERVATION: THE VISIBLE CITY

A major part of urban analysis is the visual perception of a city; i.e. the recording and interpretation of visually perceived urban structures. We generally perceive a city without recognizing the mechanisms of perception, yet the task of describing a city is inseparable from the act of perceiving it.

Perception is a learned communication process. In this context, perception means the understanding of visual information. We can only understand phenomena that we have already perceived. This involves generalizing from one or more recurring characteristics, previously learned by and therefore known to the viewer. Individual characteristics may vary or change over time, but the general characteristics must remain constant for the phenomenon to be recognized.

Communication structures our environment. When we accept names for things, we are unconsciously using a pattern to perceive our surroundings. This is also a condition for being able to perceive. In this sense, the human sensory organs do not transport all stimuli to the brain. Instead, like a filter, they classify, selecting from the full range of available sensory stimuli in a data-compression operation.

\\Note:
Aesthetics (from the Greek aísthesis: perception) originally meant the study of beauty. Aesthetics in the scientific sense includes all mechanisms of human perception.

Fig.5:
The structure of our perception is based on language

The process of determining names and terms is not learned discretely, but is inscribed into a culture in the form of language and perpetuated by the collective memory. The structure of our perception is based upon language. Language-based categories determine the differences (a selection process) and connections (a construction process) that we perceive, evaluate and interpret in our surroundings. This is largely an unconscious process, making it difficult for us to believe that different people perceive their surroundings differently, depending on their experience and habits.
> Fig. 5

Many of the terms we use group together very different urban features into a single category. This is like the way we perceive a human face. Rather than remembering all the individual features, what we notice is the overall impression created by the combination of those features. In a similar way, when we perceive a city, we compare it to many patterns with which we are familiar. We expect a town square, a historical church or town hall in the center of a city, for instance. In actual fact, this cognitive pattern is an idealized composite seldom seen in reality, where town centers are made up of many different phenomena, but we use our standard, familiar pattern to pick up on those parts of the overall composition that conform to the schema. Those parts that do not fit the pattern are suppressed by the perception process. If these elements form the dominant impression, then we do not perceive the city center, for instance, as being a city center. > Figs 6 and 7

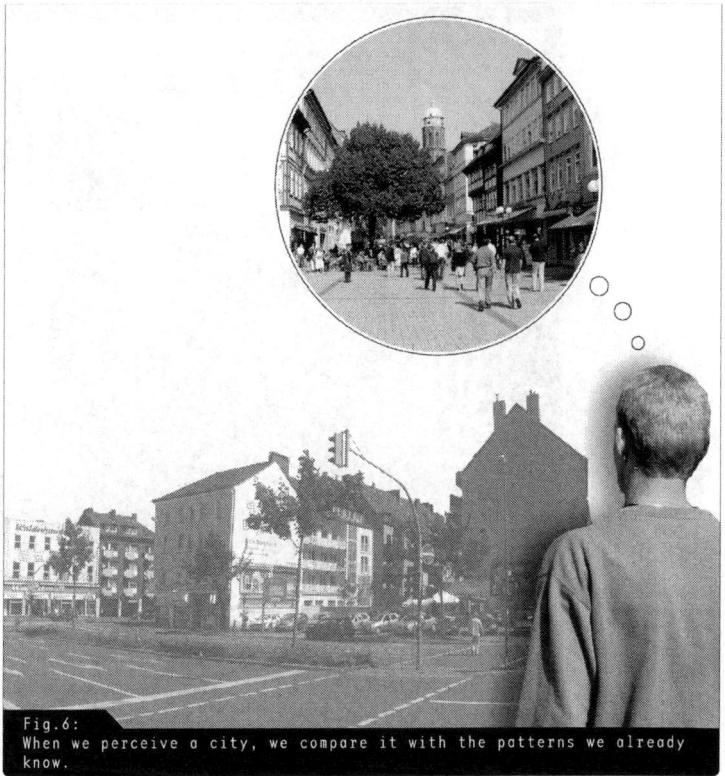

Fig.6:
When we perceive a city, we compare it with the patterns we already know.

It follows that we recognize only those phenomena that can be abstracted, based on what we already know. Or, to put it another way, we perceive only the phenomena we expect. As perception is a cultural phenomenon, urban structures may be perceived very differently by people with different cultural backgrounds.

Urban analysis may confirm familiar perception patterns by underscoring certain phenomena as being particularly typical. However, it can also help people to recognize new phenomena or connections by explaining the rules governing perception.

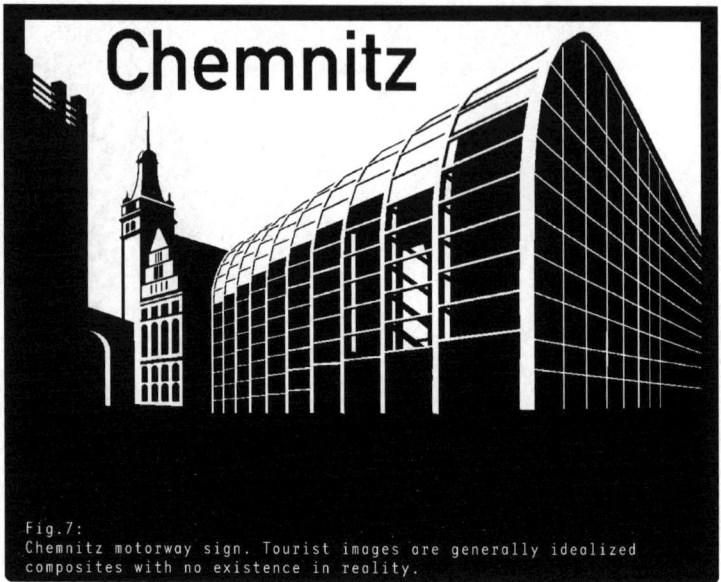

Fig.7:
Chemnitz motorway sign. Tourist images are generally idealized composites with no existence in reality.

SCIENTIFIC OBSERVATION: THE INVISIBLE CITY

A host of individual phenomena interact to produce our aesthetic perception of a city. Scientific methods, on the other hand, analyze individual aspects, with no claim to providing a general explanation of the city as a phenomenon. The smaller the area under consideration is, the more precisely it can be described. Scientists use this method to explain phenomena that lie outside the structure of perception or are not perceptible by the senses. While perception unites factors in an overall impression, science deconstructs the overall impression to discover the individual factors. For instance, characteristics of a district can be traced back to demographic features of its population without these phenomena being tangibly perceptible. › Fig. 8

Defined values

Scientific urban analysis is based on assessing and interpreting defined values. The resulting measurements are usually interpreted by comparing them with measurements from other survey areas or timeframes. Analyzing statistics, such as the age or employment status of the population, is a typical example. The interaction of all active values in an urban space cannot be scientifically investigated. Individual aspects, however, can be described precisely, as exact measurements and comparative values are available.

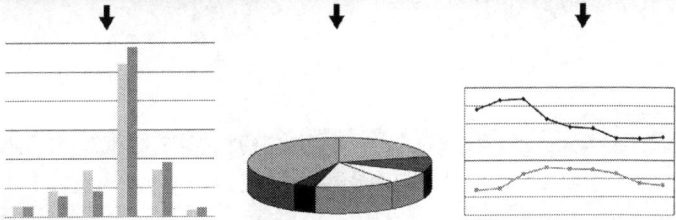

Fig.8:
Science explains phenomena that lie outside the structure of perception.

INTEGRATIVE OBSERVATION: THE LOGICAL CITY
Morphological description

Urban morphology describes the spatial characteristics of city structures and explains the conditions and causes behind them, including, for instance, the cultural, political or topographic context. Urban morphology, therefore, often follows a cause-and-effect pattern, with particular circumstances creating specific spatial structures.

Urban morphology involves intensive study of historical urban ground plans, i.e. the distribution of streets and building plots, usually based on a fundamental distinction between planned and non-planned or organic cities. Other criteria include the density and distribution of the buildings, the situation and hierarchy of streets and squares, the proportion of open spaces to built-up areas, and the placement of significant

\\Note:
Morphology is the science of shapes and forms.
Urban morphology describes the form of cities
and residential estates and their processes of
formation.

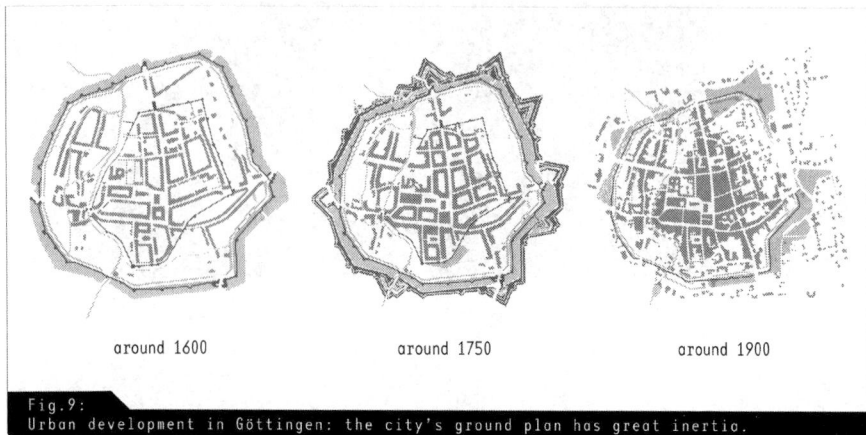

around 1600   around 1750   around 1900

Fig.9:
Urban development in Göttingen: the city's ground plan has great inertia.

buildings in the city's ground plan. Individual structural features are often studied, including:

_ Development structure: individual buildings, block buildings and rows
_ Access structure: streets, commons, squares and bridges
_ Open space structure: open space and bodies of water

> 🖪

While individual buildings have a comparatively short lifespan, the ground plans of cities are generally very slow to change. Street plots in particular possess great inertia. Historical cities therefore contain the traces of many different eras. While street plans often date from the founding of the city, most buildings will have been replaced or modified many times, so that the remains of many different eras, all of which arose under different circumstances, are jumbled together within a city. This means that morphological urban analysis may provide many explanations within a single city. › Fig. 9

🖪
\\ Note:
Analysis of urban morphology can reveal the logic of an area by investigating the conditions and causes of its characteristics.

Fig.10:
Construction styles can help us work out the age of a building.

Analysis of building style

Analysis of building style compares the formal composition of buildings and urban structures from different eras (historical styles), regions (regional styles), or cultural movements. In rare cases, individual architects or municipal building officers originate a style that characterizes a city. Style is always created by the culturally active minority, so that rather than being urban space's primary shaping force it is an intellectual, formal expression of traditional building experience.

The study of style does not necessarily explain the cause of each style. Even so, particular worldviews or construction technologies can help to elucidate a style, giving the subject applications beyond "decoration".

Analyzing styles often helps us to work out the age of buildings or urban structures. Not every era, however, has its own distinctive style. Often, several different styles were used simultaneously, or there were long transitional phases between stylistic eras. Determining style and date of origin is often made more difficult by the redesign of historical buildings and changes to the city's ground plan, so that stylistic analysis often relies on detailed historical analysis. › Fig. 10

Analysis of functional patterns

Most methods of analysis describe urban structures as a consequence of other circumstances, reasoning that social, climatic, economic or cultural conditions bring about particular spatial urban structures.

Fig.11:
Even if designs differ …

Fig.12:
… the functional pattern of a building
entrance is readable to all visitors.

Analysis of functional patterns assumes that this works both ways: as well as social and cultural activities creating spatial structures, spatial structures have implications for social and cultural behavior. One might say that spatial structures act as a catalyst for the social activities of the urban space's users.

Readability
of spatial
structures

We are not born knowing the social behavior appropriate to given spatial structures. It is learned, like a language, and is tied to a cultural context. Any changes to spatial structures must therefore preserve their readability if the relationship between place and behavior is not to be disrupted.

One of the key functional pattern schemas of a city designates public and private spaces. A varied repertoire of spatial structural characteristics encodes public or private city space. For instance, a square with a central monument encircled by shops declares itself to be public. › Figs 11 and 12

The association between social behavior patterns and certain spatial structures are very long-lived, but even so, new functional patterns can arise. New functional patterns usually arise by analogy with already familiar systems.

Functional patterns may be tied to particular social or cultural milieus, so that spatial codes can only be read by members of a certain social class or cultural group. As a consequence, members of different groups may have a different idea of the social behavior appropriate to a certain spatial structure, causing misunderstandings that produce social tensions. Recognizing functional patterns requires intensive study of the space and its use.

Spatial experience and emotional attachment

Physical and spatial sensations are both bodily and social experiences, making them fundamental to all human experience. All spatial experiences refer back to these fundamental experiences. Buildings are associated with people, and ensembles of buildings represent the relationships of people with each other. Associations are stored as experiences, and awaken positive or negative feelings.

Spatial structures are perceived in an emotional context, and spatial situations are associated with lifestyle. A place may be remembered, for instance, because of a particularly pleasant event, with an unimposing place remaining in the memory because a future life partner was seen there for the first time.

Individual and collective experiences

Many of these emotional links are based on individual experience, and therefore play no part in urban analysis. However, spatial situations can also stir emotions in society at large or in certain social groups. In some cases, a spatial situation can even produce opposite emotions in two different social groups, thereby creating conflicts.

Such places usually stand out from their surroundings. In general they are either particularly exposed places, like viewing platforms, or boundaries between two very different spatial structures, like riverbanks. Such locations are often spatially marked, creating a symbolic emphasis. In urban analysis, emotionally significant places must be recognized and documented. › Fig. 13

\\Example:
This system also applies to the ground plans of buildings, making the furnishings and locations of rooms with a specific purpose readable even to people using them for the first time. Rooms are full of codes that link to certain behaviors. Violating the rules by disregarding this code is socially prohibited.

\\Example:
Memorials relating to the two World Wars can produce very contradictory emotions in people, because they arouse both sorrow and guilt.

Fig.13:
Marking an emotionally significant place

Micro and macro levels

Cities are perceived on different levels. People usually have detailed knowledge of the area surrounding their home. In parallel with this, they incorporate the area into wider contexts without reference to detailed knowledge of these wider surroundings. This network is oriented on structural elements such as green corridors, the course of rivers or major traffic routes. Conspicuous places provide points of reference within it. Every location is therefore integrated into spatial structures on both the micro and the macro levels. Neither of these levels is autonomous. Instead, the two different levels overlap within any specific location.

IMPLEMENTATION

PREPARATION AND PLANNING PHASES

In theory, a surveyed area contains endless volumes of information, allowing a corresponding amount of analysis to be done. In practice, certain systems have become established within which analysis takes place. These systems can be divided up under the headings history, housing geography, and social structure. > Chapter Methods of analysis

Defining the subject matter

There are two reasons for conducting an urban analysis:

_ To reveal the overall picture: This means using urban analysis to obtain an overall picture of the area of investigation or of a whole urban area. Observed mistakes often trigger an urban analysis. These may include mass vacancy, poor building maintenance or major population fluctuations. The aim of the urban analysis is to replace suspicion with a definite diagnosis. In this case, several methods will be used on the area under investigation.

_ To provide an answer to a specific issue: This means analyzing specific features – for instance, the condition of existing residential buildings, the population's social situation, or the urban planning compatibility of a prospective building project. In this case, the evaluation must provide conclusions on the initial question.

Urban analyses are not formal or standardized procedures. This makes defining the subject matter before beginning an analysis indispensable. It is important to choose the right methods of analysis based on the reason for conducting the analysis. Failing to appropriately define the

\\ Note:
Housing geography describes both the spatial structure and the internal functions of the area under investigation. The most important elements are the functional structure, access structure, open space and green structure, and construction and development structure.

Fig.14:
Overall urban analysis: the smallest
spatial unit is a building plot.

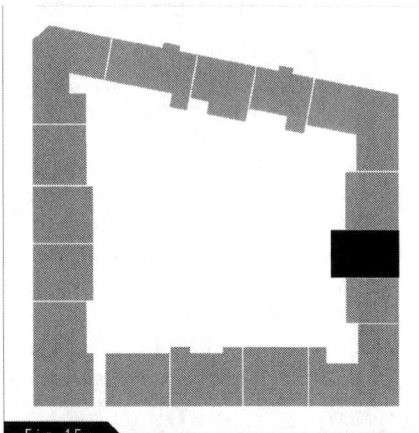

Fig.15:
Analysis of a building plot: the
smallest spatial unit is a building.

subject matter can lead either to inadequate results or to collecting information that is of no use to the urban analysis. A rough impression of the
situation within the area of analysis is therefore needed before the subject
matter can be established.

Scale and spatial demarcation

Any information gained from an analysis is subject to scale. For instance, information relating to individual buildings may be relevant to a
single building plot but useless for analyzing the whole city. It is not that
small-scale analysis will always give an inaccurate picture of the area of
investigation; it is simply that to provide useful results it must be seen in
the wider city context. An overload of too much large-area information – a
"data graveyard" – can distort findings on the area's relationship with the
wider context. Compressing the data obtained into key points is an important part of urban analysis.

The smallest unit for a whole-city urban analysis is a building plot
or neighborhood (scale 1:20,000 to 1:5000). This means that any analyzed
characteristics must be seen in relation to this spatial area. Where the area
under investigation is a single neighborhood, the smallest units are generally buildings and plots. Analyses involving a survey of only a few characteristics generally include a higher degree of detail. > Figs 14 and 15

Fig.16:
When analyzing the area of investigation, nearby areas must also be taken into account.

Scale and degree of detail depend on how far the area under investigation extends; i.e. the larger the scale, the smaller the area of investigation. However, no planning area is autonomous. It will have many connections and reciprocal relationships with the surrounding areas, the city as a whole, and even the wider region. Any urban analysis therefore has to look beyond the area under investigation and take into account its integration into the city's structure. › Figs 16–18

Continuation and monitoring

Continuing the analysis involves building on the investigation at particular intervals. Continuations may involve the same methods of analysis, or expand on or modify them. This can lead to the analyzed facts experiencing a change in emphasis.

\\Note:
In this case, "building plot" describes a housing development enclosed by roads or other structural elements.

\\Note:
An area of investigation's spatial integration situation may have a positive impact — in the case of nearby recreational facilities, for instance — but it may also create conflicts, as in the case of an industrial concern emitting fumes near a residential area.

Fig.17:
Phenomena relevant to different areas
are active in any specific location: …

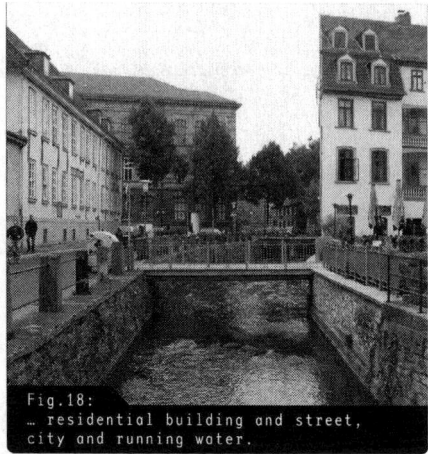

Fig.18:
… residential building and street,
city and running water.

Monitoring in an urban analysis context means long-term observation of the area under investigation, usually by surveying a constant set of values at regular intervals, comparing the results, and comparing with other areas of investigation, or comparing several areas of investigation with regard to certain values. Typical surveyed values include the number of inhabitants and the demographic and social population characteristics.

Working with other specialist planners

It is often impossible for a single specialist planner to conduct an urban analysis, as many aspects require specialized knowledge. This means

\\Note:
So that results can be compared, monitoring demands clearly establish assessment factors and working methods. It therefore generally uses only standardized values. Comparability of the surveyed areas should also be checked.

that, when preparing and setting an agenda for an urban analysis, the factors analyzed and the expertise needed must be clearly set out. In practice this involves setting up a working team or dividing the work between public planning authorities and specialist planners. Urban planners, architects and housing geographers usually supervise the main stages of an urban analysis, investigating land use, transport, housing and construction structures. Specialized analysis may require, for instance, transport system planners or historians. Open space and ecological issues are generally handled by open space planners or landscape architects. Biologists, meteorologists and geographers make specialized contributions on the open space and ecological situation. If analyzing the social structure necessitates extensive surveys or statistical work, social scientists will have to be included. Participation initiatives will require moderators.

OBTAINING AND PROCESSING DATA

Urban analysis involves obtaining extensive amounts of data. Permission to access these materials must be agreed or arranged as early as the availability and capability of the relevant authority will permit.

Urban analysis involves collecting and processing very different form of data:

_ Data in paper form, such as printed reports and plans
_ Digital data, such as planning documents, text and image files
_ Information gained from personally visiting the site › Chapter Working in the area under analysis
_ Information from conversations with third parties › Chapter Working in the area under analysis

\\ Tip:
When establishing a working group, areas of research should be clearly demarcated to avoid redundancy. If necessary, the chronological sequence of the analysis should be firmly established, as the results produced by one specialist planner may have implications for another's evaluation.

\\ Tip:
Before beginning an urban analysis, it should be established whether there will be any charge for providing or preparing necessary data materials, and who should be liable for this.

If an urban analysis is to be published, someone must check the copyright status of the materials used, which should be identified in the publication. Data protection laws apply to the publication of personal data. The data protection authorities responsible should be contacted so that rules for dealing with these materials can be agreed.

Historical information

> 🖊

Historical maps, publications on the city's past and chronicles are important reference works for a historical analysis. It is generally a good idea to talk to city archives and libraries. In smaller cities, these services are often provided by honorary appointees. They will however only provide a collection of historical materials such as photos and local publications, rather than scientifically prepared information. The evaluation of such historical raw materials is very labor-intensive, and is usually only required in the case of a scientific investigation.

Maps as a basis

Scale maps are an essential practical tool in any urban analysis. Maps may provide information or be a medium for contextualizing the different results of analysis. There are no national or international standards for maps as a basis for urban analysis, meaning that the availability and technical quality of the materials have to be checked at the beginning of the analysis. Because they may use different projections, not all sets of maps are compatible. This means that they cannot always be used within the same document. Maps for urban analysis use are usually produced and distributed by state bodies.

Topographical map

A topographical map shows geographical features and spatial objects on the earth's surface. The contour lines in particular make this kind of map useful for an urban analysis. Scales of between 1:5000 and 1:50,000 are used for urban analysis purposes. Maps are generally available as

🖊
\\ Note:
If the area under investigation contains objects of cultural significance (buildings or locations) or listed ground-level features, then authorities responsible for listed structures will usually be able to provide detailed records and experience.

🖊
\\ Note:
The display style of a topographical map makes it easily accessible to the general public. It can therefore be included in public projects without modifying the display.

Fig.19:
Topographical map

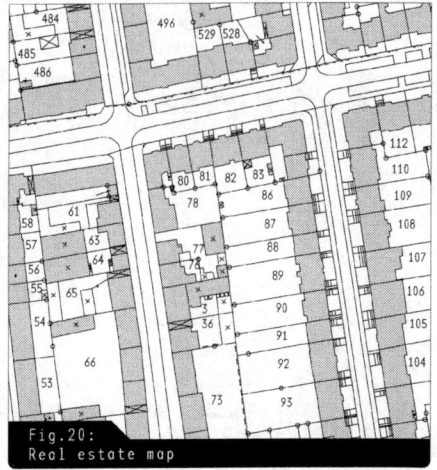

Fig.20:
Real estate map

> 🖿

digital data grids. To some extent, groups of objects or geographic features can be provided as layers, each printed with one set of data. › Fig. 19

The real estate map is used as a basis for the geodata, enabling numerous connections with other specialist data. The real estate map is the visual part of the land register, showing the ownership of all parcels of land.

The real estate map shows spatial and topographical features as well as plots of land. Its inclusion of extensive information on objects and land plots makes the real estate map particularly interesting to urban analysts. Real estate maps note particular attributes of plots and objects, e.g. land use or number of stories in buildings. Identical or similar objects are organized using the precise definitions in an object catalogue. These are arranged into separate layers, sorting the map's content by subject. The real estate map is also part of the land register map information system – i.e. the database contains extensive data on the area in addition to the real estate map's visual information. › Fig. 20

Real estate maps are kept digitally by land registry offices and can be edited by others if converted into vector files. Before referring to the real estate map, it should be ascertained what object and plot-related information it can provide. Conversely, it may be necessary to narrow down a large amount of available data.

> 🖿

The usual file formats for transfer into a CAD (Computer Aided Design) system are *.DXF (Drawing Interchange Format) or *.DWG (Drawing, a file format from the Autodesk product range). If files from the land registry information system are to be transferred, other data formats must be used.

Aerial photograph

Aerial photographs are available taken vertically or at an angle. Non-distorted true to scale vertical images – called orthophotos – are often used, as these are comparable to other map materials and can therefore be used in combination with them. Together with real estate maps, aerial photographs can provide important information on object coordinates. In particular, this helps to pinpoint green infrastructure. Aerial photographs are also an effective way of illustrating analysis results because they can be easily understood by lay people. › Fig. 21

Detailed maps

Depending on the area of investigation, an urban analysis may require large-scale detailed maps. For instance, utility companies will be able to provide maps showing underground supply lines.

Geoinformation systems

Geoinformation systems (GIS systems) are used to manage and visualize spatial data. Many evaluations can be made based on them. With GIS systems, the emphasis is generally on storing extensive bodies of digital data in the form of databanks and making connections between them through evaluation rather than expressing information in maps. The reference values for these connections are spatial features, which are recorded graphically in geometric form – as with a CAD system. These drawings are usually based on standard maps such as the topographical map.

Conducting an urban analysis using a GIS system requires extensive prior research. The availability of compatible sets of data is very important. Data transfers from other systems can lead to data being lost during formatting, necessitating time-consuming manual inputting of data. The

---

\\ Note:
If the real estate map covers a large area, it may be a good idea to exchange test data (extracts) before submitting a final set of data.

\\ Note:
Real estate maps are technical maps with limited relevance to lay people. They must be extensively revised graphically before they can be used to illustrate analysis results.

Fig.21:
Vertical aerial photograph (ortho-
photo)

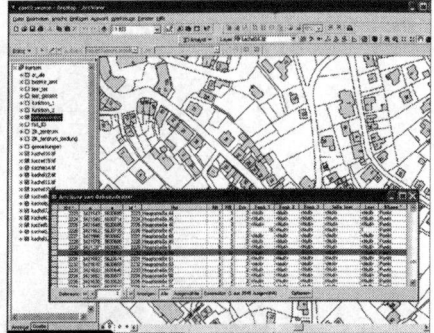

Fig.22:
GIS systems relate spatial features to
a databank.

commonest providers of data records are public bodies, but private concerns are increasingly providing spatial data records. One's own survey
results can be incorporated into the GIS system as well as external data
records. > Fig. 22

Field computers   Field computers allow standardized survey results to be collated
digitally onsite. The results being entered can be assigned to the correct
location by clicking on a touchscreen showing a map or by matching with
coordinates calculated by an integrated GPS (Global Positioning System)
receiver. A GIS system evaluates the submitted data. > Fig. 23

Existing plans and analyses
Areas under investigation have often been previously subjected to
planning and analysis. These materials must be evaluated in the light of
certain restrictions, and of other information. Formal planning works
set out regulations on, for instance, the type and extent of land use for

\\Note:
The use of GIS systems for long-term area
observation (monitoring) is particularly
interesting, as the inclusion of the most
up-to-date sets of data makes it relatively
easy to create new evaluations and evaluative
maps.

31

Fig.23:
An example of a field computer

construction, the situation of any feature for which the immediate surroundings must be kept clear, or the extent of flood-risk areas. Some-
times specialist assessments are available, analyzing particularly important factors in the area of investigation and its surroundings. These
include:

_ Ecological specialist articles containing information on existing
  valuable green infrastructure > Chapter Analysis of open space and green
  structure
_ Sound protection reports dealing with noise pollution – existing or
  anticipated in connection with planned projects
_ Transport reports analyzing existing traffic pressures and the capacity of existing transport facilities, networks and systems > Chapter Analysis of transport structure
_ Retail reports analyzing the retail structures of the survey area
  > Chapter Analysis of land use structure

Any relevant information from the available reports will be incorporated into the urban analysis. Its interactions with other factors and its
implications for future urban planning will be investigated. Depending on
its importance, this information will be represented in a scale map.

> 🖉

Data and
statistics on
residents

In general, administrative authorities are required to carry out regular analyses of the population structure. This involves analyzing the whole
population of districts or statistical survey areas under certain headings.
The area of investigation for an urban analysis is generally not identical with any of these statistical survey areas, making the possibilities
for applying the data to a specific investigation area limited. To make an

Fig.24:
Formal planning texts contain information important to an urban analysis.

evaluation relevant to the area of investigation, a new census area has to be marked out.

This demarcation must be done by administrative departments, because external experts do not have access to data on residents. In order to satisfy data protection requirements, data sufficient for building up a small-area picture can be provided only after extensive agreements have been reached between urban analysts and the authorities responsible.

When analyzing the population structure, it is particularly useful to supplement surveys restricted to specific reference dates with studies of developments over time, and compare them with other survey areas, making it essential to have access to suitable comparative data.

\\Note:
In practice, all existing plans and analyses are requested from the relevant authorities in the early stages of an urban analysis. Many planning texts can be downloaded from the Internet (see Fig. 24).

\\Note:
The term "census area" describes a continuous area for which particular characteristics such as the average age of residents or the number of homes can be assigned.

WORKING IN THE AREA UNDER ANALYSIS
Taking an inventory in the field

Field work in the area of investigation is an essential part of urban analysis. Most analysis results can only be obtained by recording the existing situation on-site. Spatial qualities and atmospheric impressions only become apparent to an urban analyst visiting the area of investigation. It may be necessary to view the area at different times of day, or on different days of the week.

You will need to have suitable documents prepared for textually and graphically recording the information gained from the site visit. The volume of information recorded and the nature of the spaces involved will vary depending on the dimensions and degree of detail of the analysis. The documents must be appropriate to the volume of information and the spatial relationships involved. Unsuitable documents with excessive notes can turn the subsequent evaluation into a labor-intensive piece of detective work. › Fig. 25

Parties to the onsite visit will generally evaluate their own notes. More extensive surveys, however, involve assembling an editing team. In this case, a standard way of listing analysis results must be agreed upon to preclude subsequent misunderstandings. Standardized questionnaires are particularly suited to this.

Topographic or real estate maps are used to record the onsite visit. Different scales are used depending on the dimensions and degree of detail of the analysis:

_ For analyzing the urban integration situation, topographical maps or aerial photographs with a scale of between 1:5000 and 1:20,000 are suitable.

\\ Tip:
To prevent any information gaps in the subsequent evaluation, the site visit must be preceded by thorough preparation. Such an error may result in time-consuming extra research, particularly if long traveling distances are involved.

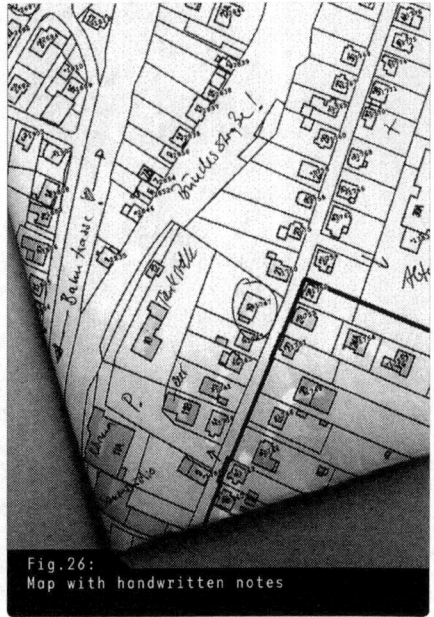

Fig.25:
Log book for recording building-
related characteristics

Fig.26:
Map with handwritten notes

_ For recording housing geography features, real estate maps with a scale of 1:1000 are generally used. When preparing these maps, any layers that are not relevant to the site visit should be left out.

_ For recording plot or building level data, real estate maps with a scale of 1:500 are used. Larger-scale maps tend to result in paper formats that are too large to handle easily in the field. It may be a good idea to combine the map and the survey sheet, with the map allowing individual objects to be pinpointed and the survey sheet allowing analysis results to be listed. The survey sheet will be created as a standardized questionnaire, to prevent any survey criteria from being overlooked during the inventory. > Fig. 26

Cameras are generally used to document the area of investigation. It is often a good idea to note the places where photos were taken on a map, to allow the photographs to be put in a spatial context during subsequent analysis.

Citizen participation, contact with local operators

Residents have extensive detailed knowledge of their own neighborhood. Others active in the area, such as business people and those who run local initiatives, will also have knowledge of local conditions and how they affect people. These sources of information should be included in any analysis.

Regulated contact with residents can be achieved using a series of methods. This is useful for assessing the impact on people within the area of investigation of all the factors and conflicts involved, rather than for gaining detailed information on the locality.

Residents'
meeting

A residents' meeting is a frequently used method of citizen participation, usually with little preparation. In practice, residents' meetings where no specific action is discussed have a low perceived level of impact on the residents and are therefore not well attended. Organizing a residents' meeting is therefore only recommended where both specific measures and general plans of action are to be discussed. It is also a good idea to present residents with the first urban analysis evaluations, to provide an object and structure for the subsequent discussion. As some residents will avoid the public exposure of meetings, this method will not produce a representative sample of residents. The meeting place chosen should always have a low associated inhibition level – i.e. should be within the investigated area.

Individual
public
consultation

Preparing, carrying out and evaluating an individual public consultation is very time-consuming, and they are therefore only included in urban analyses that require an extensive social structure analysis. Individual consultations and their evaluation require the assistance of social scientists.

Individual consultations may take the form of narrative interviews or standardized questionnaires. Standardized processes make it easier

\\ Tip:
A visit to the site will generally result in spontaneous contacts with residents. Such meetings should not be avoided – careful inquiries may provide useful information.

to evaluate the results. Mass-mailed questionnaires generally have a low response rate. They do not provide a representative cross section of the population structure either.

Specialist
consultations
Conducting several specialist consultations allows a large body of location-specific specialist knowledge to be gathered in a relatively short time. Specialist consultations are discussion groups focusing on specific subjects, with specialists in the relevant field invited to attend. Participants generally include specialist representatives from the relevant authorities plus those active in the area under discussion. These may be people working for welfare organizations, voluntary workers or representatives of local entrepreneurship.

METHODS OF ANALYSIS

Urban analysis of a particular area may include investigations of one or more issues, e.g. different land uses and transport organization in the area of investigation. As already stated, these sectoral analyses are simply abstract, partial representations of the reality. In practice, however, they simplify the analysis process, as the complex interplay of all values cannot be recorded or presented. At the same time, any urban analysis should note as many interactions between these sectoral analyses as possible. Put simply, urban analyses begin by dismantling the area under investigation like a motor, and then use the knowledge of the individual parts thus gained plus their interactions to put it back together again.

Urban analysis is not formally regulated, and so there are unlimited ways of carrying it out. For this reason, only the sectoral analysis types most commonly used in practical town planning are included here.

HISTORICAL ANALYSIS

Historical analysis inquires into the causative events for the creation and composition of the analyzed area, or significant influences on it. Events in recent history may therefore also be significant. Historical analysis is not limited to identifying the stylistic roots of existing urban structures. It records and interprets events that had implications for the spatial development or the area of investigation. › Figs 27 and 28

There are two possible fields of reference for a historical analysis:

_ The area of investigation as a whole: This involves evaluating which historical events and circumstances have influenced the overall structure of the area of investigation.
_ The area of investigation as the sum of its parts: This involves researching individual structural features to determine whether they are or were typical of particular developments.

Historical events include spatial measures such as the building of town fortifications or reconstruction after a fire. Many historical events, however, have implications for the area of investigation's spatial development without themselves being specific spatial measures. For instance, economic circumstances such as a boom or decline in the regional economy may have affected the area of investigation's spatial development.

Fig.27:
Göttingen's former town fortifica-
tions …

Fig.28:
… are now part of a vital recreational
area.

Recording
individual
features
Whether the area of investigation will be analyzed as a whole or on the level of separate plots is generally decided by the number of historically significant buildings. In a plot-by-plot analysis, buildings are assessed individually, and any historically relevant features documented. Buildings are usually assessed according to a standardized catalog of criteria, then – if possible – assigned to various building types. Commonly used criteria include the building's age, and the extent and state of preservation of any historical stylistic features. The next step is to mark these buildings on a map based on their features. This helps to ascertain the spatial distribution of different building types in real space and the number of groups of significant buildings. › Figs 29 and 30

Interactions
in the area of
investigation
In some rare cases, the spatial characteristics of an area of investigation can be explained by a historical event. This would be the case, for instance, for a comprehensively planned neighborhood built within a limited period of time. More usually, however, an area of investigation's spatial characteristics have been influenced by several different events and circumstances. These events will generally have occurred at different times, leading to repeated remolding of the area of investigation, and leaving traces on various places within the area of investigation (buildings from different eras). Different events may have molded a single place, with one set of circumstances overlying another (e.g. 1950s buildings within the ground plan of a medieval town). The density and distribution of analyzed features will show whether the area of investigation's development

39

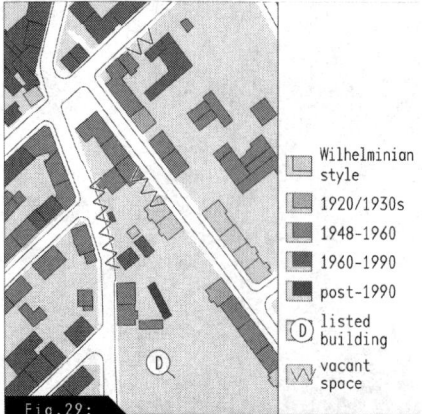

Wilhelminian style
1920/1930s
1948-1960
1960-1990
post-1990
listed building
vacant space

Fig.29:
Historical analyses either record individual features…

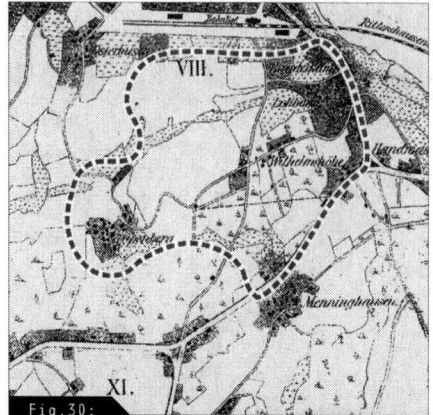

Fig.30:
… or consider the area of investigation as a unified whole.

can be traced back to a small number of historical developments and circumstances, or whether a very diverse historical development pattern lies behind its present characteristics.

› 𝟬

Context beyond the area of investigation

Historical analysis involves working out the relationship of the area of investigation to surrounding areas or to the city as a whole, to understand its context within the development of the city. It must also be ascertained whether the area of investigation is typical of the overall development of the city, or whether it has played a special role in history.

Conclusion

Thorough historical research is particularly important in areas significant to the city's planning history. However, areas initially considered to have no historical significance may reveal typical development processes in the course of a historical analysis, demonstrating aspects of the area of

𝟬
\\ Note:
When examining the area of investigation as a whole, historical maps are generally used to analyze historical development. Due to differences in scale and reproduction, these cannot usually be compared with modern-day maps.

𝟬
\\ Note:
Land use intensity often correlates with construction density, meaning that the results of the building and settlement analysis should be compared with the results of the use structure analysis.

investigation's identity and ongoing development. When interpreting historical analyses, it is important to keep to describing the facts that have clearly influenced the area of investigation's development. Discursions on general city history, e.g. succession, documentary references or town twinning arrangements, can be excluded from an urban analysis, as they generally have no influence on the properties of an area of investigation.

ANALYSIS OF LAND USE STRUCTURE

Analysis of land use structure involves investigating the real uses that exist or predominate within the area of investigation and the interrelations of different uses.

Land use intensity

Land use intensity can be described using statistical values. A ratio is created from all the area of investigation's compiled characteristics. One application of this is to calculate the proportion of surface area under different land uses by expressing the proportion of the overall surface area taken up by each individual use. However, features may also be documented in this way – the number of workplaces, for instance. Land use intensity can also be described as a density value. This involves calculating the relationship of certain features to units of surface area. One frequently used density value is population density relative to settled area.

Recording individual features

Analysis of land use structure involves determining real use relative to spatial units. A spatial unit in this context may be an open space or a building. If the facts are being recorded by area, and the area is occupied by a building, the building's function is the determining factor for its use type. For instance, if the building on the site contains apartments, then the designation will be "residential" regardless of any open space near the plot.

Analyses of whole cities or investigation areas of comparable size establish land uses for whole building plots (or neighborhoods), i.e. areas containing several plots. As these spatial units often have no homogenous land use structure, they determine the predominating use instead. District land use structure analyses are generally done plot by plot, i.e. determining the land use of every plot, or the number of plots in a particular land use area. This involves recording the surface area used for traffic and green spaces as well as built-up plots.

Analyses of land use structure may be restricted to recording building use, leaving open space use to the analysis of open space and green structure. This division of labor is particularly suitable when land use is

Residential
Mixed use
Commercial
Secondary
facilities

Fig.31:
Depending on scale and the degree of
detail required, a land use structure
analysis may be performed plot by plot ...

Residential
Mixed use
Commercial

Fig.32:
... or with reference to particular
building plots.

being recorded story by story. Such highly detailed analyses are usually only applied to small areas of investigation. › Figs 31 and 32

Land use is generally researched by visiting the site, i.e. visually. However, much information can be gained from real estate map data. In some cases, authorities may provide data on surface area use from their GIS systems. › Chapter Obtaining and processing data Another useful source of information is the directory of firms kept by some business associations. There is however a risk that the postal address of a business is within the area of investigation, but not its production sites.

The land uses present within an area of investigation can be differentiated to varying degrees depending on the degree of detail needed. Subsequent division of the results into headings and subheadings is recommended. › Fig. 33

Residential

Generalized residential functions can be differentiated by the building structures, e.g. detached, semi-detached and row housing, or multistory apartments with open or closed building forms. Special residence types – such as sheltered housing and homes – are shown separately, as they make particular requirements of local suppliers and transport connections.

Residential buildings
as part of a row

Multi-story
apartment buildings
(open building form)

Multi-story
apartment buildings
(closed building form)

Individual houses

Mixed construction
areas

Commercial
construction areas

Urban wasteland

Residential and
access roads

Other routes/roads

Parking areas

Retail of
everyday goods

Fig.33:
Land use structure analysis map with key

Commercial

Commercial land uses are differentiated according to the degree of disruption they cause – i.e. their compatibility with residential land use:

- _ Industrial trade creates the greatest degree of disruption, and is therefore confined to planning-regulated industrial districts.
- _ Trade encompasses all commercial land uses. These exist mainly in planning-regulated industrial districts. They include manufacturing, haulage and storage, wholesale, city depots and craft enterprises.
- _ Non-disruptive trade includes types of business compatible with residential land use. Most of these are in the service or retail sectors.

43

The output of the service sector is non-material, and depends on direct contact between people. Service sector trade can therefore be differentiated according to the degree of dealings with the public:

_ Customer-oriented service sector trade describes facilities that regularly deal with the public, such as travel agents or leisure services
_ Non-customer-oriented service sector trade includes offices

Service sector operations may also be suppliers, meaning that supply and non-supply service sector operations can be differentiated. Service sector operations with a supply function include doctor's surgeries and offices of postal and parcel services.

Retail describes businesses that sell products to the end user. By contrast, wholesale dealers sell to sellers.

In a district-level analysis, retail facilities for everyday essentials are additionally identified because they are important local amenities for residents. Local providers of everyday essentials should not be farther away (as the crow flies) than a 500 m radius (for a good local amenity situation) or 700 m (for a satisfactory local amenity situation) from any place of residence. Everyday essentials include groceries and everyday luxury items, drinks and health care and grooming products. Services such as doctors' surgeries and post offices may be included in the evaluation of the local amenity situation.

Mixed land use describes a combination of residential and other – usually residence-compatible – land uses. Different combinations – for instance, residential and restaurants, residential and offices – can be differentiated, depending on the degree of detail of the urban analysis.

🗋
\\ Note:
Retail can be subdivided according to many different criteria, for instance the marketing type (self-service, supermarket, specialist shop, specialist market etc.), centrality, area served, branch group, and shopping area.

Identifying different degrees of combination to show whether residence or commerce is the dominant land use is also common.

Public amenities

Public amenities are public facilities open to all – particularly kindergartens and daycare facilities, churches, cultural facilities and authorities. Open-air facilities such as play areas and cemeteries may also be recorded as public amenities.

Leisure and social facilities may be public amenities, but may also be run by commercial concerns. Depending on their nature – public or private – they may be evaluated as public amenities or customer-oriented service-sector commerce.

Hotel and catering

Catering includes all facilities offering catering services, e.g. restaurants, snack bars, cafes, liquor stores and ice-cream parlors. Hotels include guesthouses with service.

Entertainment venues

Facilities intended for entertainment and relaxation, such as discos and amusement arcades, are entertainment venues. As they generally create a fairly high level of disruption, they are treated as a separate category in urban analysis.

Plant

Plant includes facilities or buildings used to regulate and maintain various types of supply for the urban area or to control particular commercial processes. These include gas regulator stations, transformer stations, and sewage plants. Many of these facilities are surrounded by protection zones, i.e. a certain distance must be maintained between the plant and surrounding land uses.

Transport

Transport structure analysis differentiates between different types of road or different financial responsibilities for roads. In either case, land use structure analysis can also provide findings on transport facilities.

Transport-related constructions, such as rail stations and bus terminals, may be recorded as such, or listed as public amenities.

Open-space uses

Important open-space uses, generally included in analysis of open space and green structure, can be incorporated into the analysis of land use structure instead. They include parks and play areas, which can be classified as public amenities.

Vacancy

Vacancy is a functional deficiency. Mass vacancy generally indicates that an area is unattractive, usually due to urban planning disruptions or mistakes. It is difficult to establish and prove vacancy, partly because it is not always obvious to the eye and partly because it takes long-term observation to know how long buildings have been vacant.

Vacancy in residential and commercial buildings are differentiated, as these often have different causes and require different measures. It is also wise to differentiate between part of a building and a whole building standing vacant. › Fig. 34

Other land uses
Unusual uses that occur only once or rarely within the area of investigation are generally identified and labeled individually in the evaluation plan, to avoid lengthy and confusing keys on maps.

Garages, shelters and similar facilities can be recorded as secondary facilities.

Granulation
The main task of land use analysis is to investigate subdivisions with comparable land use types and thereby determine granulation. Recording these main areas is particularly important for identifying any disruptions between different land use areas. Land uses each produce their own level of disruption, but also an entitlement to protection. High land use gradients – neighborhoods with very different protection requirements and degrees of disruption – generally lead to conflicts, while shallower use gradients – neighborhoods with similar protection requirements and degrees of disruption – have low conflict potentials.

Context beyond the area of investigation
When investigating subdivisions with comparable land use types, the areas adjoining the area of investigation must also be considered, as these can also be the cause of disruptions or protection requirements.

\\ Note:
Aside from judging by appearances, vacancy can be deduced from buildings' electricity use. This method involves the energy supply companies, and is therefore rarely practical.

Fig.34:
Vacancy may indicate that a structure is not fit for purpose.

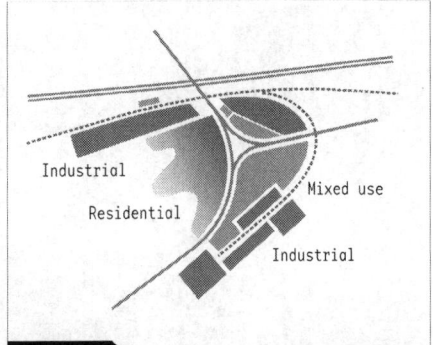

Fig.35:
Abstract sketch illustrating the integration situation

Industrial
Residential
Mixed use
Industrial

An area's granulation reflects land use distribution and the area of investigation's land use balance. This relationship also expresses the purpose and centrality of the area of investigation in the broader urban context. Categories are generally based on the different planning area types, e.g. › Fig. 35

_ Purely residential area
_ Residential area with a limited number of residence-compatible workplaces
_ Mixed area
_ Inner city and city center
_ Commercial area
_ Industrial area
_ Special area

Conclusion

During the final evaluation, the purpose of the area of investigation must be taken into account, to determine which findings harmonize with that purpose, and which contradict it. › Chapter Interpretation and illustration A central task of land use structure analysis is to reveal any land use conflicts. A distinction must be made between the cause, e.g. incompatible combinations, and the effect, e.g. increased vacancies. › Fig. 36

Fig.36:
Mixed land use structure: appropriate
to area of investigation or source of
land use conflicts?

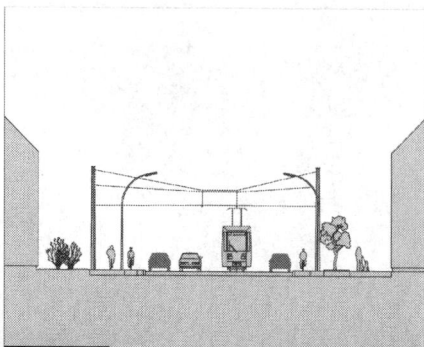

Fig.37:
The spatial features of a road can
be represented by standardized cross
sections.

ANALYSIS OF TRANSPORT STRUCTURE

Transport structure analysis investigates existing transport facilities in the area under investigation and the connection and integration of the area of investigation into transport networks. In this context, roads are public spaces as well as transport constructions, making them critical to perception of the area of investigation. Road design should therefore also be the object of urban analysis. Often characteristic features can be identified for each road. It makes sense to record the standard of improvements to each part of the road network. These represent both the capacity and user-friendliness of the various areas of road, e.g. › Fig. 37

_ Proportion of road width to adjacent buildings
_ Division of road space into driving lanes and side areas
_ Type, shape and size of road spaces
_ Road furniture and surface materials
_ Lighting apparatus and lighting types

A highly detailed analysis of transport structure will require a specialist engineer. Transport reports suitable for urban analysis purposes may already be available. As well as analyzing types of transport (passenger or freight), purpose of transport (shopping, professional or leisure journeys etc.), and distances (source traffic, destination traffic, internal and transit traffic), these are likely to contain data on traffic pressures.

Transport can be analyzed according to many different structural features. In urban analysis, it is mainly differentiated according to the type of transportation used, i.e. motorized traffic, local public transport and foot and cycle traffic.

Motorized traffic

Moving traffic

The first step in analyzing moving traffic is to rank the roads in the area of investigation according to their capacity, e.g. › Fig. 38

_ Expressway / trunk road
_ Main traffic route
_ Connecting road
_ Main collector road
_ District access road
_ Other road significant for transport
_ Agricultural and forestry routes

This kind of analysis determines the area of investigation's road transport network structure. In most cases, all its roads form a hierarchical network; i.e. the traffic in the network is unevenly distributed. Road types are based on a kind of division of labor. Heavily used roads concentrate the traffic, while moderate to low-use roads distribute it. Parts of the network can be assigned functions based on their traffic pressure: connection (heavy load), access (moderate load), and stopping (low pressure). Deviations from this schema tend to lead to conflicts.

› ◻

Dormant traffic

Parked vehicles (and non-functional vehicles) are described as dormant traffic. An urban analysis generally notes the provision and uptake density of public parking spaces. Public parking spaces may be found on

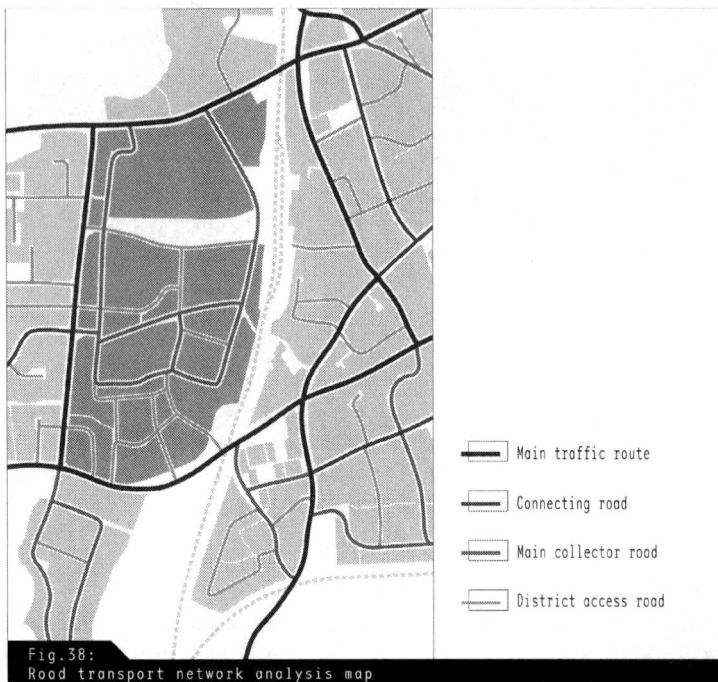

Fig.38:
Road transport network analysis map

Main traffic route

Connecting road

Main collector road

District access road

public roads or in designated collection facilities (ground-level car parks, underground garages, or multi-story car parks). These are managed in different ways, ranging from non-paying parking space, to timed parking, to a charge for parking. › Fig. 39

Different land uses in an area of investigation cause different parking space requirements. This means that the spatial location of parking spaces in relation to different land uses is of interest as well as parking space provision.

Context beyond the area of investigation

Transport structure analysis involves the area of investigation's internal access network as well as how the area of investigation is integrated into the road traffic network of the city or region as a whole. This means determining the shape of any transport network within which the area of investigation is contained – a radial or ring system, for instance. An area of investigation generally has traffic networks that cover different units of distance overlapping within it. Long-range transport systems are a direct

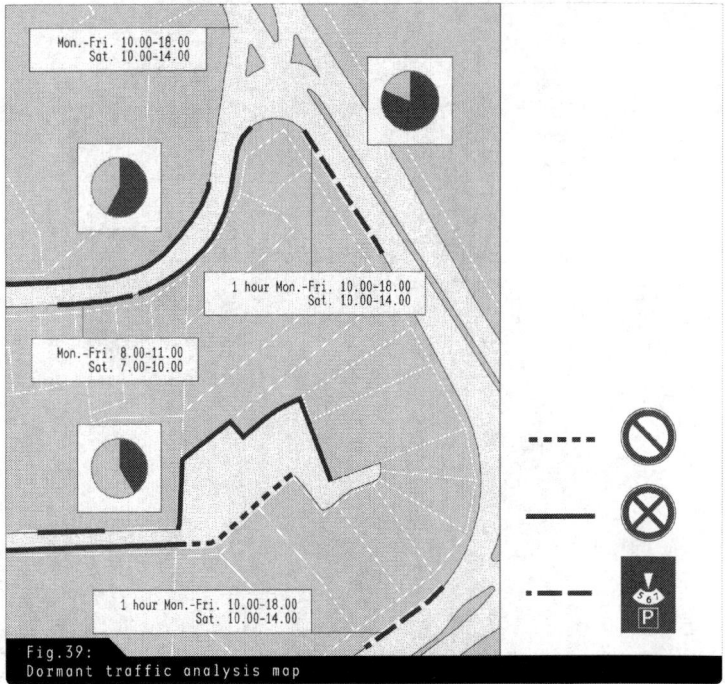

Fig.39:
Dormant traffic analysis map

means of transport to beyond the area of investigation, but also burden it with traffic headed elsewhere, with the associated fumes and noise emissions. Depending on where these networks and functions come together spatially, they may lead to disruptions within the area of investigation.
› Fig. 40

\\Note:
Uptake intensity is investigated within a representative time period. This involves calculating the number of occupied parking spaces in relation to the total number. An occupancy rate graph showing the degree of use over the course of a day or week can also be compiled.

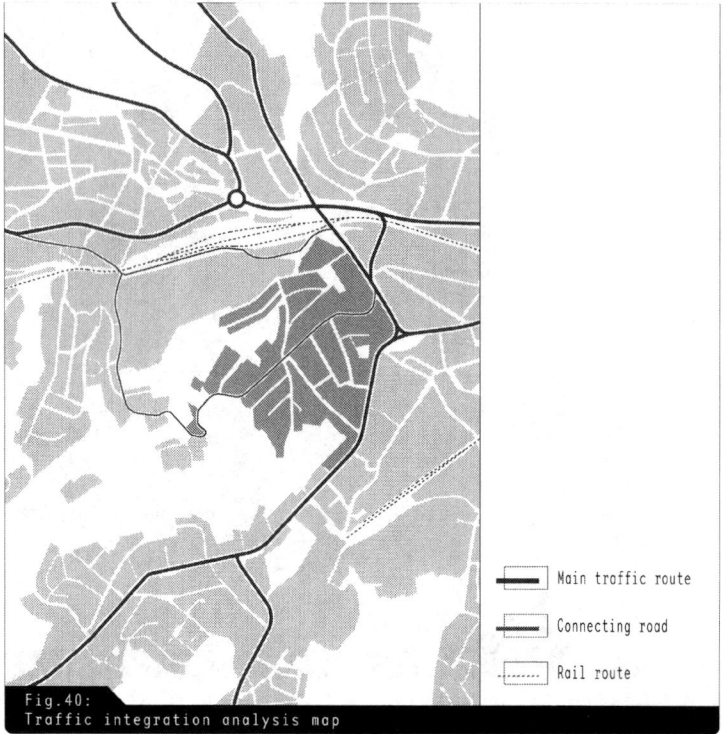

Fig.40:
Traffic integration analysis map

	Main traffic route
	Connecting road
	Rail route

Local public transport

Local public transport describes short-distance passenger services accessible to all. These generally follow routes and schedules. Analysis of local public transport investigates the availability of and service at stops, and the network's structure. A distinction is made between different local public transport systems:

_ Railbound systems (regional trains, suburban trains, city railways, subways, trams)
_ Scheduled bus system
_ Other systems (e.g. on-call shared taxi, citizens' bus)

The provision of stops is an important factor. A maximum distance of 300 m from the relevant residential and working areas (as the crow flies) represents good network access, while a maximum of 500 m represents

Fig.41:
Analysis map of short-range public transportation

satisfactory network access. This analysis should take into account topo-graphical conditions, existing pedestrian facilities (e.g. pedestrian cross-ings) as well as spatial barriers that require a detour. The level of improve-ments (e.g. lifts serving rail platforms, high curbs in relation to low floor technology) also affects accessibility.

The quality of local public transport depends significantly on fre-quent service at stops, as well as the network speed. Punctuality, regular scheduled connections, long running hours and mechanisms for speeding up public transport (e.g. bus lanes and automated priority for local public transport at traffic lights) are important measures of quality when analyz-ing local public transport. › Fig. 41

Context beyond the area of investigation

The area of investigation may be contained within a centrally ori-ented or decentralized network. Centralized networks have a single node where passengers can change for other route lines or transport systems. Decentralized networks have several nodes. Many networks fall somewhere between these two systems. The upshot of all this is that when assessing local public transport the area of investigation's spatial proximity to these nodes is as important as the number of stops it contains.

Foot and cycle traffic

Depending on the degree of provision in the area of investigation, pedestrian and cycling facilities may be independent elements or simply parts of the network.

Analyses of the footpath system only include independently managed footpaths, with walking routes along the sides of roads only included in very detailed transport structure analyses.

An analysis of cycle traffic will generally record all cycling facilities, and distinguish between cycle paths along the side of roads, cycling zones (marked routes within general traffic lanes), independent cycle paths, and cycle roads etc.

The improvement standard of the pedestrian and cycle network and its safety systems are the major factors determining connection quality between different parts of the area of investigation and its surroundings. For this reason, the nodes of all facilities for conducting cycle traffic should be recorded, not just the safe crossings (e.g. crossings at intersections and other ground-level crossings). Areas that are particularly inconvenient or dangerous should also be named. All land uses within the area of investigation that represent points of origin or destinations for various forms of transport should also be recorded. › Fig. 42

Context beyond the area of investigation

Footpaths and cycle paths connecting the area of investigation with its surroundings or with neighboring districts are of particular interest to urban analysis. For analyses of whole cities, or of areas of investigation of comparable size, it is a good idea to divide the pedestrian and cycle routes into leisure and everyday networks. While leisure networks generally represent access to regional recreation facilities, everyday networks are based on reaching the foci of urban activity as quickly as possible.

Concluding observations on transport structure analysis

Different land uses have different access needs – i.e. they require transport systems and facilities of different dimensions. The task is to

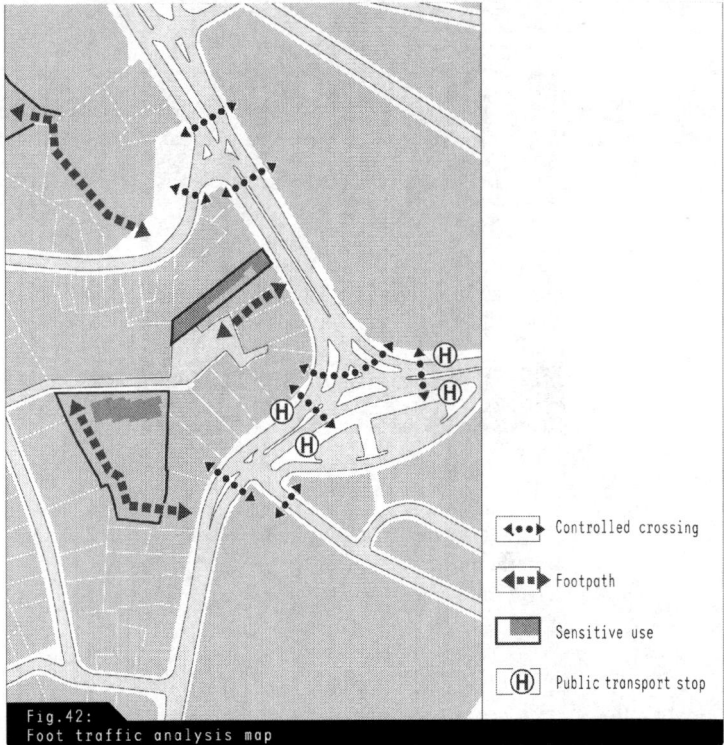

Fig.42:
Foot traffic analysis map

◀●●▶	Controlled crossing
◀■■▶	Footpath
▨	Sensitive use
Ⓗ	Public transport stop

identify the varying degrees of provision and any disruptions, along with the cost and quality of access provision. Conversely, all the land uses in an area of investigation have varying rights to protection from dust and noise emissions. High pressures from motorized traffic tend to impact on the quality of nearby housing and the quality of time spent in an urban space. They also act as spatial barriers, dividing the area of investigation. At the same time, high traffic pressures on roads can represent centers of urban life. A concluding evaluation should weigh up all these significant factors.

The different kinds of transport also make competing demands on urban space. One transport network's high standard of development may be detrimental to the functioning of another. For instance, plentiful bus lanes may be convenient for users of local public transport, but impact on the capacity for individual motor transport.

Fig.43:
Roads must be evaluated in the light of many different interacting factors.

All networks should be assessed for whether they support spatial orientation by being readily comprehensible and easy to remember. Good orientation increases network users' comfort and sense of security. › Fig. 43

ANALYSIS OF OPEN SPACE AND GREEN STRUCTURE

Analysis of open space and green structure analysis has two main tasks:

1. To represent open-space infrastructure elements and their right to protection under species protection and ecology law. Urban analysts do not generally perform their own ecological analyses, instead taking any information relevant to the area of analysis' spatial development from external specialist reports.
2. To investigate the use and aesthetic quality of open space and the interactions between open space and the housing structure. The spatial qualities of the green and open space structure are personally surveyed and evaluated by urban analysts.

Space-related features

Like a land use structure analysis, open space and green structure analysis records all real use of open spaces. Unlike land use structure analysis, the object of the process is the use of the open spaces on the plots involved. The type of open space represented by each individual space is recorded. As the subject matter overlaps with land use structure, this labor-intensive process is often left out of urban analysis, in which case

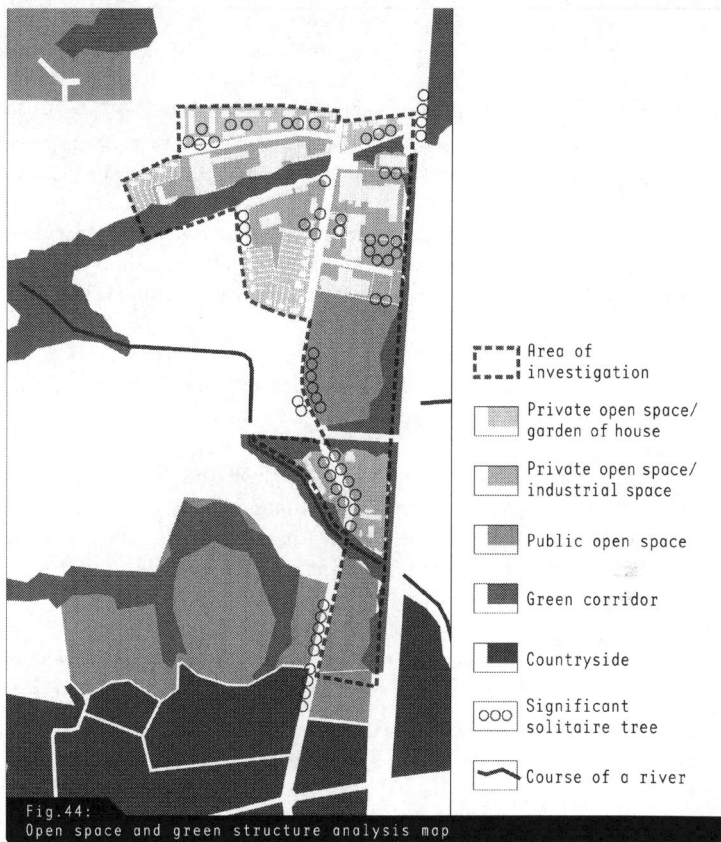

Fig.44:
Open space and green structure analysis map

Legend:
- Area of investigation
- Private open space/ garden of house
- Private open space/ industrial space
- Public open space
- Green corridor
- Countryside
- Significant solitaire tree
- Course of a river

the main features of analyzing open space and green structure are simply differentiated. These spaces are differentiated primarily by their ownership, recording > Fig. 44

- Public open spaces, e.g. public parks, fairgrounds, waterside promenades, bodies of water
- Public open spaces associated with certain facilities, e.g. school and kindergarten playgrounds, zoos
- Semi-public open spaces, e.g. the inner courtyard of an apartment complex, an office building's open spaces
- Private open spaces, e.g. gardens, company grounds

Landscapes are generally only differentiated by open space type in the case of a whole-city analysis. The following are recorded:

_ Cultivated space (e.g. farmland, pasture, vineyards)
_ Woods (e.g. protected woodland, recreational woodland)
_ Other spaces (e.g. moorland, marsh, bodies of water)

As well as recording open spaces, open space and green structure analysis also records natural structuring features that have a particular effect on the space around them. These structural features include:

_ Notable solitaire trees and groups of trees
_ Linear structures such as avenues of trees or hedges
_ Green corridors

Spatial features of the green and open space structure are hard to designate on a map using a standardized key, as so many of them are highly individual. When documenting analysis results, it is therefore a good idea to use spatial sketch plans or abstract icons. › **Fig. 45**

The first step in analyzing open and green structure is to evaluate the intrinsic value of individual features. Elements that have particular significance in one or more of the following four categories are evaluated:

_ Historical significance: significant garden architecture or cultural history elements (e.g. historical parks and elements of the cultural landscape)
_ Spatial experience: spatial qualities, such as the interplay of planting and open space or the cumulative effect of open space and topography
_ Recreational value: areas with special recreational functions (e.g. parks and municipal woods near housing areas)
_ Functional value: uses of open space and green structure are important as part of surface area-related land use recording
_ Protection value: elements of the open space and green structure protected by law

An inventory of all protection zones and protected elements taken from the relevant planning documents and specialist reports must be incorporated into the urban analysis:

_ Protected zones where particular land uses are prohibited: Protection of animal and plant life, landscape characteristics (e.g.

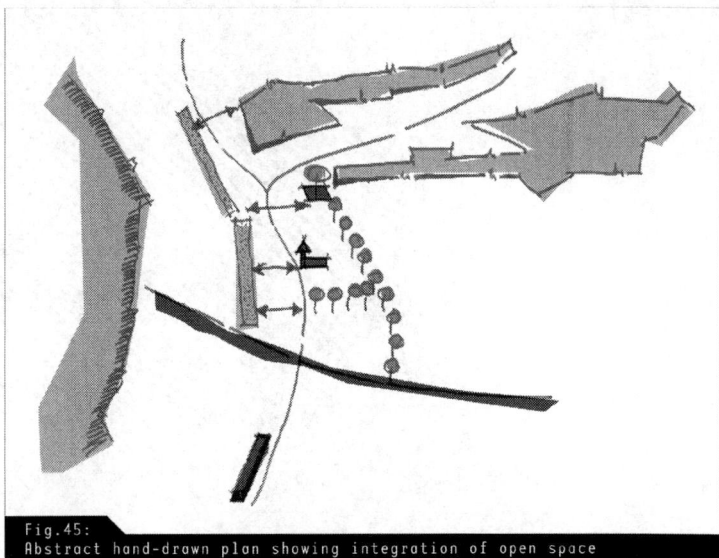

Fig.45:
Abstract hand-drawn plan showing integration of open space

conservation areas and protected landscapes), and protection of drinking water resources and fresh air corridors

_ <u>Protected infrastructure elements</u>: Protection of individual elements for animal and plant life, such as solitaire trees, hedges or banks

For an urban analysis, all protected zones and protected elements relevant to the area of investigation's urban development are pinpointed on a scale map. > Fig. 46

\\Note:
The recreational value of open spaces varies according to land use intensity. Open spaces close to settlements often have many visitors, whereas continuous countryside has proportionally fewer visitors.

Fig.46:
Protection zone analysis map

Interactions in the area under investigation

The interaction of open space and housing structure is the most interesting part of open space analysis. The proportional surface area and the spatial distribution of each type of open space can be determined from an inventory of open spaces. Calculating this enables us to determine the degree of provision of specific recreational spaces for the area under investigation. This is generally according to the population of the area of investigation.

Residential estates with many small units on the edge of cities generally have houses with gardens, reducing the need for public open spaces. Closed, heavily built-up inner-city estates, on the other hand, have a greater need for public recreational space.

Context beyond the area of investigation

When analyzing open space and green structure, the area of investigation's integration into its surroundings and into the city as a whole must be recorded. Of particular interest are the spatial connections between the individual green spaces and green infrastructure elements, and how the area of investigation's open spaces connect with their surroundings.

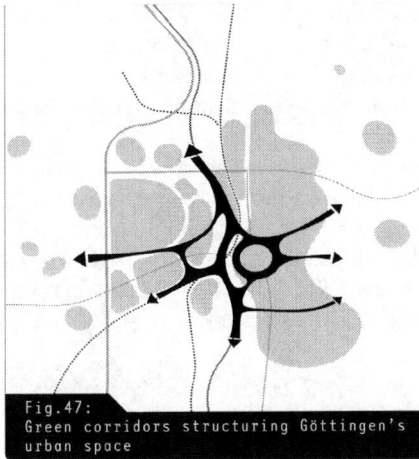

Fig.47:
Green corridors structuring Göttingen's
urban space

Open space and green structure may have characteristic geometries, for instance a ring or central corridor system. These geometries often overlap – e.g. in a concentric open space system. ⟩ Fig. 47

Small connective elements and narrow green corridors may also be important in this context. Analyzing the integration situation also reveals whether open spaces outside the area of investigation are providing its residents with recreational space. The following are recorded:

_ The integration of housing into the landscape, connections between housing and open space, access to recreational spaces near housing
_ Green connective elements, e.g. avenues or grass verges
_ The interplay of open space and topography, e.g. green riverbanks or planted embankments
_ The spatial effect of green infrastructure elements, e.g. solitaires, rows or groups
_ Landscape profile, e.g. open or closed landscape

The qualities of many open space and green structures are only revealed in interaction. These networks are often incomplete, with gaps or spatial barriers. For purposes of general planning orientation, the connecting lines and corridors that only partially exist when the analysis is made should be recorded.

ANALYSIS OF CONSTRUCTION AND HOUSING STRUCTURE

Construction and housing structure analysis investigates the area of investigation's spatial composition, describing development characteristics and the construction development-open space relationship.

Recording individual features

Construction and housing analyses can be conducted to varying degrees of detail. For a plot-by-plot analysis, the structural features of every individual building are recorded based on a standardized catalogue and entered on a map. This degree of detail is necessary in urban analyses focusing on historically significant features, for which the following structural features might be recorded: › Fig. 48

- Number of stories and building size (degree of construction saturation)
- Building type (e.g. grouped houses, block, ribbon development)
- Building form (e.g. closed or open building form)
- Roof shape (e.g. pitched roof, hipped roof, flat roof)
- Building orientation (e.g. with front or gable facing the street)
- Access system (e.g. apartments with stairway access, block with outdoor-corridor access, block with inner access)
- Added stylistic elements (e.g. ornamentation, beam decoration)
- Building construction (e.g. solid construction or with a facade)
- Age, construction era
- Ownership situation (form of rental or ownership, diversified holdings or large-scale operator)

Granulation

The plot-by-plot analysis is used to identify subdivisions containing similar structural features, separating the area of investigation into sectors according to housing structure and original construction phase. This is generally done on the basis of building size and type and the age of the development. It is easiest to do this for areas with homogenous urban planning, i.e. subdivisions with unified development. Typical development structures include:

- Closed block development
- Open ribbon development
- Compact ground-level construction
- Freestanding apartment complex
- Family home development

› 📖

It is harder to differentiate areas with a non-uniform, apparently chaotic structure, which often make up most of the investigation area. In

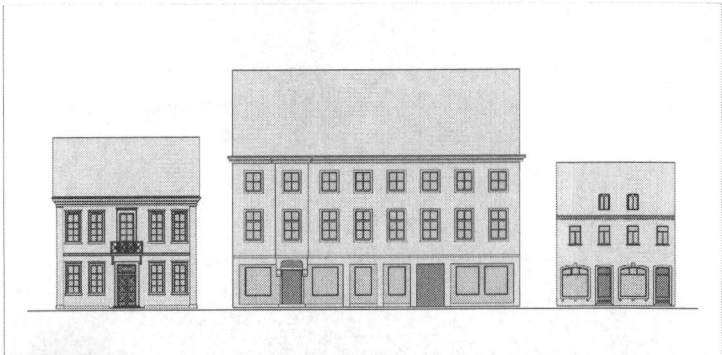

Fig.48:
An individual building assessment often records the main facade's
significant features.

this case, the degree of uniformity is used as the differentiating factor.
Areas may, for instance, be listed as:

_ Compact block structure dating from the turn of the 20th century
_ Largely compact block structure from the turn of the 20th century,
  the 1950s and the 1960s
_ Open development structure with buildings from all periods

Construction and housing structure analysis often dispensed with
this time-consuming individual assessment of buildings. In this case,
subdivisions with similar housing structures are given at the outset. In
a city with a mixed structure, this requires relevant practical knowledge.
> Fig. 49

> \\ Note:
> Further information on urban development types
> is given in *Basics Urban Building Blocks* by
> Thorsten Bürklin and Michael Peterek, Birk-
> häuser Verlag, Basel 2008.

Fig.49:
Identifying areas of specific ages and with specific housing structures

Urban spaces are created by the interaction of numerous buildings, or rather by the interaction of buildings with open spaces. These spaces have very different characteristics, e.g. › Figs 50 and 51

_ Demarcation: spaces with closed or open character (clear demarcation or gradual transitions)
_ Geometry: Spaces formed by geometrical edges (e.g. by building lines) or by irregularly positioned buildings
_ Proportions: Relationship of buildings to open spaces, different densities and extents

Conversely, it is also necessary to identify imperfections and disruptions within the composition.

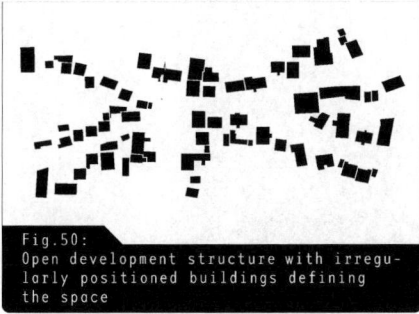

Fig.50:
Open development structure with irregularly positioned buildings defining the space

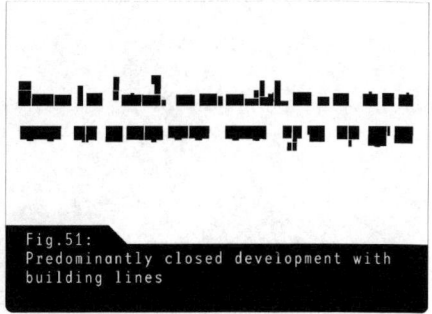

Fig.51:
Predominantly closed development with building lines

Context beyond the area of investigation

Construction and housing structures are usually influenced by spatial structures and structural elements that extend beyond the area of investigation. This often reveals the intrinsic logic of a housing structure and assists navigation within the city. Typical relationships include:

_ Topography and housing structure: specific housing structures built on topographically different areas (e.g. large buildings on level ground, smaller-unit housing structures on less level ground)
_ Orientation on a central space: certain housing characteristics increasing as a center is approached (e.g. increased housing density in the center of town relative to the periphery).

Exceptions to these rules are often described as disruptive, but may also be particularly exposed buildings or spatial elements that have important roles as landmarks.

Conclusion

Analyzing the construction and housing structure provides information on the function of the area of investigation. This can best be done by comparing the construction and housing structure analysis with the results of the land use structure analysis. A dense inner-city neighborhood with mixed land use structures generally takes on supply functions for other city neighborhoods. Due to their dense ground plan, however, these also place greater demand on public recreational spaces, which may need to be provided outside the area of investigation.

ANALYSIS OF SOCIAL SPACE

Social space analysis looks at the interaction of space with population structure, identifying urban areas within which population and spatial structure features are relatively constant. Social space analyses are

Fig.52:
Analyzing demographic features: the sizes of the statistical survey areas are based on neighborhoods.

generally conducted by social scientists, with urban analysts comparing the most important findings from their reports with other urban analysis results. This process creates insights useful to urban social planning, highlighting the spatial and topical key points for urban development and for the general social interest. › Fig. 52

Working methods   There are no universally established social space methods of analysis. Two different ways of posing the initial question, and two different working methods, are possible:

_ What is the spread of particular features of the population?
_ Which specific population features exist within a particular spatial area?

Referring to
existing
evaluations

In practice, city administrators continuously observe the population structure and its development within the boundaries of the city. As a rule, the individual statistical areas are based on administrative boundaries. An

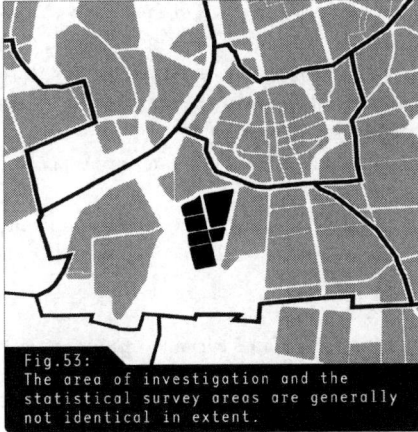

Fig.53:
The area of investigation and the
statistical survey areas are generally
not identical in extent.

area of investigation for urban analysis, however, is generally not identical with an existing statistical area. Referring to existing observations will therefore produce only an inexact picture of the situation within the area of investigation. › Fig. 53 An analysis tailored to the area of investigation involves requesting data from the relevant census authority. This means that data availability, technical implementation, and data protection should be discussed at the outset.

Type and extent
of surveyed
characteristics

The simplest form of social analysis involves recording the population. By recording the population on a series of reference dates, population development can be recorded.

› 🗋

By calculating the number of inhabitants relative to spatial units, specific densities can be determined – the housing density (number of inhabitants per hectare of occupied space) or the occupancy rate (average number of residents per home).

🗋
\\ Note:
There are many possible evaluation methods for
surveying and evaluating population data (see
Appendix, Literature).

🗋
\\ Note:
Alongside population development, the influx
and exodus relative to the net population level
(fluctuation) can give a picture of an area of
investigation's population mobility.

A number of features of the area of investigation's population can be assessed, depending on the investigative approach and the capacity of the responsible census authority. Relevant population features might include age, sex, marital status, number of children, births, deaths, nationalities, arrivals and departures, and demand on welfare services. Spatial features might include the number and size of homes or the age of buildings.

Long-term analysis is needed to reveal developmental trends. These features should therefore be recorded on several separate reference dates.

Spatial
dimensions
of surveyed
features

The unequal distribution of various urban population groups is reflected at the micro level in individual streets or building sites. If some parts of the investigated area have significantly different urban planning structures, a more fine-grained social space analysis will be needed to define the interactions between social and spatial features. This means that two different approaches are possible:

1. To relate all features to the area of investigation as a unified survey area and then compare the results with other survey areas.
2. To divide the area of investigation up spatially based on the surveyed features – i.e. to determine which features are present in which sections of the area of investigation.

> 🗎

Comparative
data

Comparing the situation within the area of investigation with other survey areas (e.g. the district, the whole city, the region) usually means acquiring data from higher-level institutions. When using different data sources, there is a risk that the recorded data were not collected using the same methods, thus making the comparison invalid.

🗎

\\ Note:
When referring to data from the responsible census authorities, the spatial area for which data on various features can be provided should be ascertained at the outset. Data protection for data on a small spatial area is particularly sensitive – the smaller the surveyed space, the greater the danger of surveyed features being traced back to individual people (see Fig. 54).

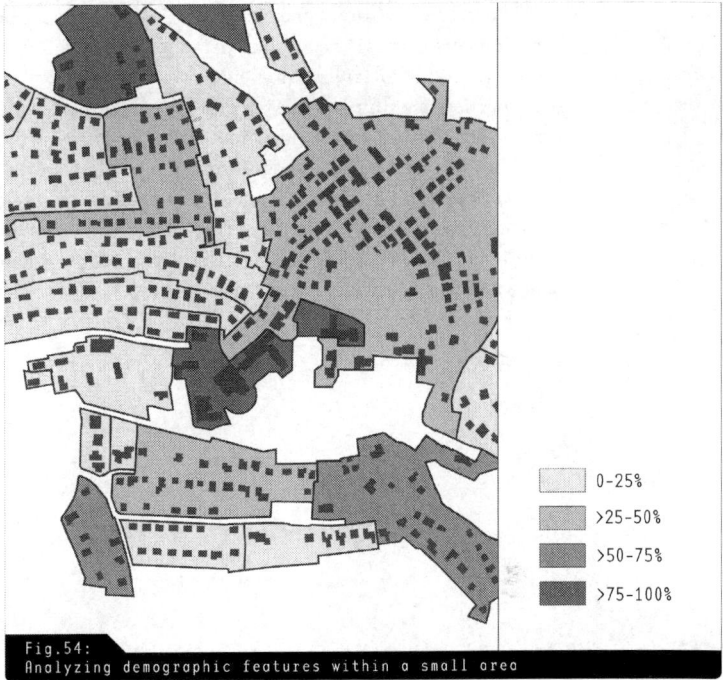

Fig.54:
Analyzing demographic features within a small area

Legend:
- 0-25%
- >25-50%
- >50-75%
- >75-100%

Conclusion

The survey will generally be used to draw conclusions about the social situation of people living in the area of investigation and the way the social situation there relates to the spatial surroundings. Reading straightforward meanings into the findings risks a false interpretation, as individual features provide limited material for substantiating such a conclusion. For instance, high population fluctuation may indicate low population stability and therefore the residents' low identification with their neighborhood. In the case of a neighborhood largely occupied by students, however, there would be nothing unusual about this. Any interpretation must therefore take into account a number of interacting features, and factor in the surveyed area's intended function. It is a particularly good idea to compare the results with those of the construction and housing structure analysis and the land use structure analysis. The expert knowledge of residents and people active in the area should also be incorporated.

> Chapter Working in the area under analysis

Social segregation, the spatial drifting apart of the population structure, runs contrary to the guiding principle of a socially mixed population structure. Social differentiation, however, means that there will always be neighborhoods with particular social milieus or classes, i.e. of people living similar lifestyles. This development is not necessarily problematic, as social milieus can make access to certain social networks easier. The spatial concentration of social extremes, however, can create ghettoes, signaling a mass presence of negative extremes in a single neighborhood and the isolation of inhabitants from social services and modernization. Any social space analysis must recognize these areas.

INTERPRETATION AND ILLUSTRATION

Initial evaluations are generally carried out for each sectoral analysis before considering the interactions between the different categories.

It is comparatively easy to evaluate an individual sectoral analysis, as comparisons can be made based on standardized values. An evaluation will usually be carried out under two headings: structure and function.

INTERACTION OF THE RESULTS OF SECTORAL ANALYSIS

A city, i.e. an entity perceived as a city is created by interacting spatial, functional and social factors. At the same time, too many competing claims on urban space are the cause of many urban dysfunctions and conflicts. Determining the difference between acceptable negative impact of one factor and another and actual conflict is an important part of urban analysis. It is not possible to record all the interactions, either for a whole city or for a single part of the city, and so the part of an urban analysis that deals with interactions, like the part that deals with sectoral factors, represents only an abstract model of the city.

Sectoral
analysis

Sectoral analysis evaluation generally follows a pattern of cause and effect: what is the cause of a spatial circumstance, and what are the consequences? Many evaluations, however, depend on the interactions between sectoral factors. For instance, a transport structure analysis may reveal that the capacity of a road needs to be increased. Once other factors have been taken into account – e.g. the location of the road within a residential area – it may not be possible to satisfy this requirement.

Comparing
sectoral
analyses

Certain needs and thereby the over- or under-provision of particular functions in the area of investigation can be determined by comparing sectoral analyses. This usually means including areas outside the area of investigation. Typical factors determining need are:

_ Development density and provision of recreational spaces
_ Population and provision of housing-associated facilities, e.g. facilities providing for everyday needs

CONCLUDING EVALUATION

Function of
the area of
investigation

The concluding evaluation must always refer back to the original starting point of the urban analysis. › Chapter Preparation and planning phases It

71

must also either answer the original question or give a comprehensive picture of the situation within the area of investigation. Answering the original question does not usually mean questioning the area of investigation's function. Instead this function – the investigated area's role in the city's overall planning scheme – decides the appropriateness of any specific new development. Analysts determine appropriate land uses on this basis. If, on the other hand, the analysis presents the whole situation rather than starting with a specific issue, strong points and opportunities and weak points and dangers must be presented. In this case, a new function for the area of investigation may be suggested after appraising these factors. With both types of analysis, therefore, the investigated area's function must be taken into consideration. Function in this case means the main role of the area in the wider context of the city, on the basis of which it is evaluated. The area may for instance function as a purely residential district but also be capable of providing a supply function for the whole city, if properly equipped. Its function must be determined in the context of the whole city – i.e. the features of the area of investigation must be seen in relation to the structural features of the city as a whole.

In the part of the concluding analysis dealing with urban dysfunctions, the functional and substantial weaknesses of the area of investigation can be differentiated. A problem that prevents the area of investigation from fulfilling its function is a functional weakness. If a problem with the area of investigation's material make-up is detrimental to healthy living or working conditions, or endangers people living and working there,

Strengths	Weaknesses
- good access	- gaps in local
- plenty of	services
green space	- noise pressures
- cultural diversity	- vacancies
Opportunities	**Dangers**
- a place for young	- increase in
families to live	vacancies
- conversion of empty	- damage to local
buildings	image

Fig.55:
Arrangement of positive and negative aspects in an evaluation matrix

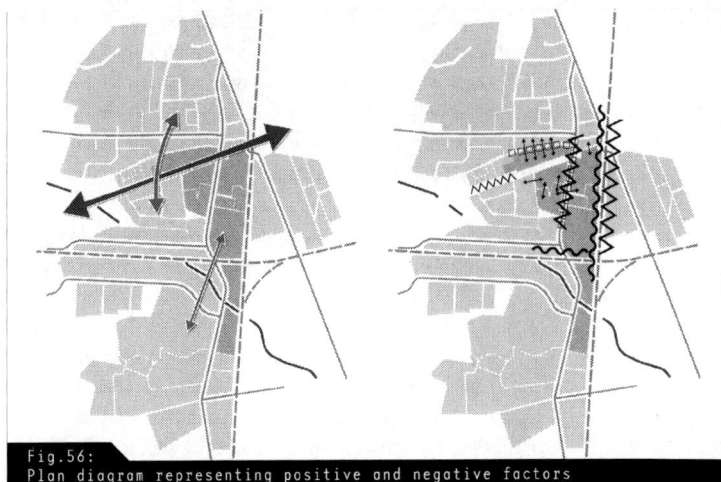

this is a <u>substantial weakness</u>. Substantial weaknesses may apply only to parts of the area of investigation.

Assessing opportunities and dangers means looking beyond the status quo, using the analysis results to identify the area of investigation's possible future development. Analysts must base these theories closely on the area of investigation's spatial tendencies in particular. Opportunities might include, for instance, converting wasteland into public green space or using empty buildings for new forms of housing. Dangers might include a continuation of negative trends or committing urban planning errors.

Evaluation
matrix and
graphic plan
representation

In practice, two methods are used to illustrate all these aspects:

1. Evaluation matrix representation, i.e. arranging positive and negative factors in a table. Aspects entered into the table can be placed in sequence to express their relative importance. › Fig. 55
2. A plan graphic representing positive and negative factors. This is particularly useful for showing the spatial relevance of all aspects. › Fig. 56

TRANSITION TO SUBSEQUENT COURSE OF ACTION

If an urban analysis gives an overall picture of the area of investigation, then weighing up all analysis results will form the basis for a future development reference model. As rigid plans of action often cannot be implemented in practice due to the many circumstances beyond planners' control, the following approach is advisable:

1. Establish a development model for the area of investigation
2. Sketch out several alternative action plans or development scenarios

Reference model
As a strategic overall goal, the reference model provides orientation for any future development steps. Any feasible individual measures or packages of measures can be realized without losing sight of this model, which is developed from analysis results according to three sets of rules:
> Fig. 57

_ Descriptive or analytical: identifying future developments based on researched trends
_ Goal-oriented or normative: determining required actions based on desired goals
_ Initial or investigative: investigating factors likely to give impetus to future developments

The goals envisaged by urban analysis can influence the area of investigation's development in three different ways:

1. Active and activating effect:
_ Project preparation, e.g. a feasibility study
_ Translation into an action plan or contingency plan
_ Identifying areas of potential
_ Acquiring financial support

2. Controlling effect:
_ Laying down rules for the event of a project's being carried out
_ "If ... then" options

3. Conservative or passive effect:
_ Preventing undesirable developments
_ Safeguarding existing areas of potential

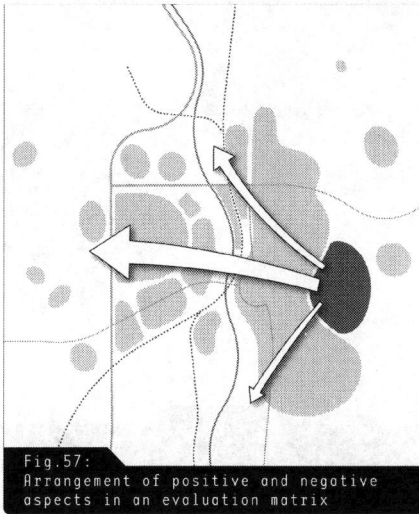

Fig.57:
Arrangement of positive and negative
aspects in an evaluation matrix

ILLUSTRATION OF THE RESULTS OF ANALYSIS

Representation
using maps

Where possible, evaluations and interpretations should be displayed in graphic plans as well as textually, in order to clarify the different aspects' spatial applications. All analysis results are entered on scale maps and given values, identified in a key. Depending on the degree of detail, it may be a good idea to divide the recorded characteristics into headings and subheadings. The key should be structured to reflect this division, with all the subcategories for a particular heading given a single color – shades of red, for instance. Representations in the plan and on the map will be based on a standardized key. Depending on the type of representation required, the outlines of buildings, plots or building sites will be used. › Fig. 58

While formal planning texts contain clear instructions for using plan symbols, urban analysis has no plan symbol standards. Preparing easily readable graphic representations of analysis results is an important part of urban analysis.

Representation-
al scale

The correct scale for representing analysis results on a map must be chosen according to the size of the area of investigation and the degree of detail in the analysis. The decisive factor here is which urban space element represents the smallest determining unit, i.e. spatial unit.

75

Fig.58:
Map key structure with headings and
subheadings

Fig.59:
Icons make important structural
features readily visible.

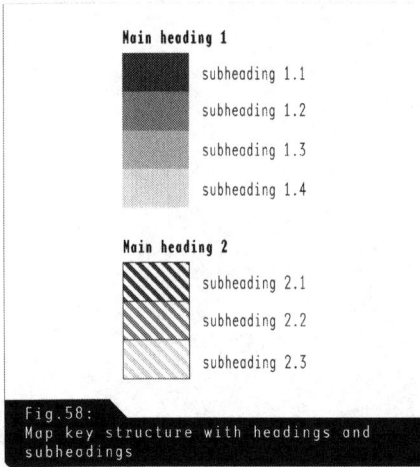

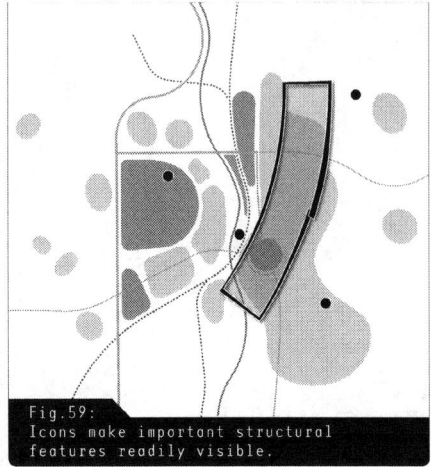

For whole-city analyses, the smallest spatial unit is generally the building site or neighborhood. Scales of between 1:20,000 and 1:5000 are used. A topographical map is likely to be used as the basis.

District-level investigations generally include information on individual buildings or plots. A suitable scale is 1:2000 or 1:1000. A real estate map is likely to be used as the basis.

Small-area investigations (e.g. individual building plots or streets) are represented on maps using scales of 1:500 to 1:200.

Sketches
and icons

Abstract sketches and icons are suitable for illustrating the most important interpretation points. Reducing these to their significant structural features enables a plastic approach to these interpretations and the spaces to which they apply. > Fig. 59

Representing
statistical
analyses

Statistical analyses can be applied to the area of investigation as a unified whole, be used to compare the area of investigation as a whole with other census areas or make statements about individual areas within the area of investigation.

Spreadsheet programs are generally used to process statistical data. These programs contain tools for representing the statistics as diagrams. If the area of investigation is being evaluated only as a statistical survey area, these tools will generally be sufficient to illustrate the evaluations.

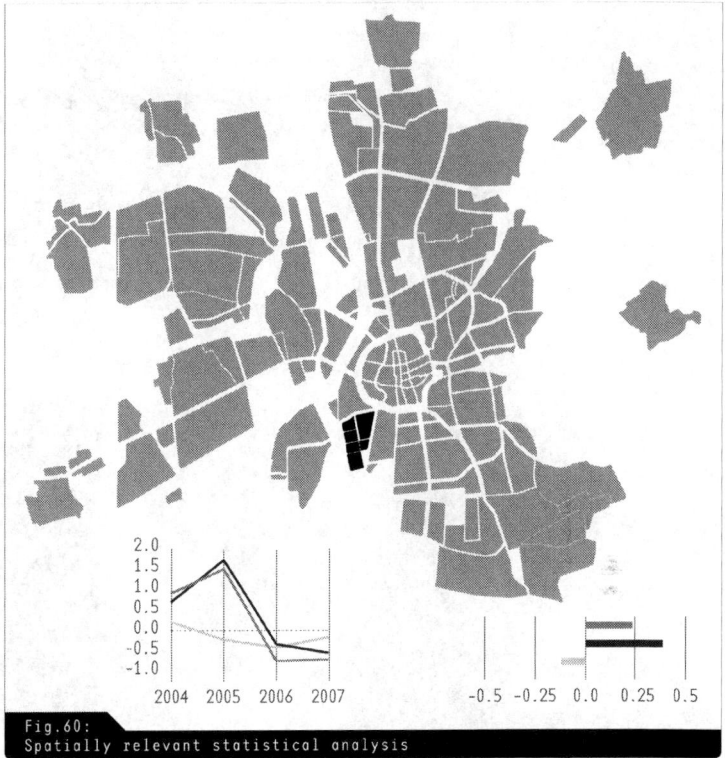

Fig.60:
Spatially relevant statistical analysis

If the area of investigation is being minutely investigated and divided into several census areas, maps showing these census areas will be needed. These evaluations are usually organized in a table of values and entered on a map according to a standardized key. › Fig. 60

Fig.61:
Deviations in perception of the area of investigation have to be taken into account in any urban analysis.

IN CONCLUSION

Urban analyses are not the sum of several individual analyses. Instead, they integrate several analyses into a single picture. The quality of an urban analysis is determined by awareness of any specific area of investigation's different interacting aspects, rather than detailed listing of the individual sectoral factors. The overall picture can only be seen when the interactions between the various sectoral factors are taken into account, and this is what ultimately determines accuracy in an urban analysis.

Extensive collecting of data on individual factors can obscure the interactions. Sectoral analyses should therefore be conducted with a view to the degree of detail the concluding evaluation will be able to handle. Digitizing spatial and population-related data has made a large amount of data comparatively quick to obtain. To prevent large, incoherent collections of data, these data must be checked for relevance prior to incorporation into the evaluation – a task requiring experience and an awareness of its specific purposes and goals.

Urban analyses create an abstract, model representation of an urban reality to serve as a basis for future urban planning measures in the area of investigation. However, as urban analyses cannot satisfactorily research certain aspects, such as emotional ties and social habits, they often contradict the perceptions of involved parties – an unavoidable fact that should be taken into account by anyone acting on the basis of an urban analysis.

Urban analyses that consult residents' wishes are the cornerstone of successful cooperations between residents, authorities, and other participants in urban life. Against a background of urban development within existing urban structures becoming more frequent, this kind of urban analysis can be a vital part of sustainable urban development.

APPENDIX

LITERATURE

Christopher Alexander, Sara Ishikawa, Murray Silverstein: A Pattern Language: Towns, Buildings, Construction, Oxford University Press, 1978

Carl Fingerhuth: Learning from China. The Tao of the City, Birkhäuser Verlag, Basel 2004

Peter Hall: Urban and Regional Planning, Taylor & Francis, 2002

Kevin Lynch: The Image of the City, MIT Press, 1960

Franz Oswald, Peter Baccini: Netzstadt. Designing the Urban, Birkhäuser Verlag, Basel 2003

Colin Rowe: Collage City, MIT Press, 1984

Aldo Rossi: The Architecture of the City, MIT Press, 1984

Thomas Sieverts: Cities Without Cities. Between Place and World, Space and Time, Town and Country, Routledge Chapman & Hall, London New York 2003

PICTURE CREDITS

Figure 7:	© kittel+partner, Dresden. Shows: Peek&Cloppenburg department store in the centre of Chemnitz
Figure 9 (left):	Graphic: G. Schwalbach. Source: The city of Göttingen in the Middle Ages, Map 2 from O. Fahlbusch: "Die Topografie der Stadt Göttingen", Göttingen 1962
Figure 9 (center):	G. Schwalbach. Source: GOETTINGA, urbs munitissima et splendida ..., an urban plan by Matthias Seutter, Augsburg (copper engraving from around 1750). Graphic design: Atelier K.-H. Fehrecke, commissioned by city of Göttingen construction authority, first published in 1971
Figure 9 (right):	Graphic: G. Schwalbach. Source: plan by municipal planning office of Göttingen 1893. Graphic design: Atelier K.-H. Fehrecke, commissioned by city of Göttingen construction authority. First published in 1971
Figures 10, 48:	Pesch & Partner: Architekten+Stadtplaner, Herdecke/Stuttgart: Gestaltungsfibel und -satzung Innenstadt Coesfeld (design manual and design statutes for inner-city Coesfeld), Herdecke 2006
Figures 19, 21:	© Geobasisdaten (spatial base data) Bundesland North Rhine-Westphalia, Bonn

THE AUTHOR

Gerrit Schwalbach, Dipl.-Ing. architect, research assistant in urban planning and history of planning in the Department of Architecture and Urban Planning at the University of Siegen, Germany

导言

城市分析与
城市规划

　　在进行城市规划时需要对所涉及的区域进行城市分析，分析的结果将为所采用的相应规划措施提供背景知识和依据。对于已有城市而言，进行城市分析——即对已有城市空间的研究——是进行城市规划的一项准备工作。

　　随着城市的发展，现代城市（尤其是在欧洲）规划在单纯数量增长方面的工作越来越少，更多的则是对城市结构方面的调整，这种转变要求对已有城市结构进行更加深入的研究。然而，存在一个基本的问题，那就是现有城市的未来发展是否能够被真实地规划出来，同时根据目前已有的条件是否能够提前分析出城市未来的发展方向。〉见图1

　　很长时间以来，城市规划一直是通过对某些特殊参数的量测和评价来确定城市空间的未来需求。从 20 世纪以来，这种方法的使用越

图1：
城镇的有机规划需要建立在完整的城市分析基础之上

83

来越得以增强。在工业革命期间，大量城市的发展经历一个几乎没有规划的扩张发展阶段，而相关专家意识到这种扩张发展在城市功能方面的不足。由于计算机的出现能够进行大量的数据处理，同时能够将城市的未来发展情况采用虚拟的方法将其进行"真实"的展现，使得相应发展的城市规划方法在20世纪60年代得到了非常重要和广泛的运用。与此同时，城市规划人员的身份也发生了改变——从最初的无责任的营造商转变成对城市进行诊断和开具处方的理性工程师。但是，所有尝试精确预测城市发展的努力均以失败告终。如今，大家认识到城市的发展过程过于复杂，根本无法完整地对一个城市的发展进行分析和推测。尽管如此，城市规划和城市分析仍然是非常重要的手段。但这些手段并不是用来规划城市发展的每一个细节，而是将城市的各个不同发展整合到一个整体计划中。由于现代城市规划越来越集中于现有城市基础结构上，所以城市分析的作用日益关键。在寻找合适的措施并采取实际行动之前，需要对已有的城市结构进行分析。城市分析并不是一个在城市规划之前进行的、二者相互独立的固定过程；城市分析和城市规划二者均是一个连续过程中的重要元素。〉见图2

城市分析　　　　城市分析是对整个城市或者城市局部区域中与城市规划相关因素的描述与分析过程。虽然说城市和城市居住者之间形成了一个相互作用的交互系统，导致无法对城市中的各个因素进行全面的评估和研究，但在城市分析过程中，仍然需要将美学、空间、社会以及经济等因素考虑在其中，并对这些因素的相互作用进行分析。因此，城市分析的任务将局限在对城市或者其中的单独区域进行抽象的模型表述。〉见图3

　　　　一般情况下，不同的人对相同的城市将会产生完全不同的感知。比如，来自某一地方社会的居住者对他们的邻里或者环境的感情联系是其他地方居住者很少具有的。相反，比如在城市设计中的失误或者过错可能对当地居住者的影响却很小。由于城市规划者与城市居民或者使用者对城市的感知和评价角度不同，导致城市规划的过程成为一个具有严重冲突的过程。最初的冲突无法避免，但若引导合理的话，这种冲突也可以转化为城市规划者与居住者加深对其关注空间理解程度的一次机会。促使居住者能够将目光拓宽到邻里之外，同时也让规划者对邻里关系和环境的社会机制有一个深入的认识。〉见图4

图2：
大量的城市单体开发组合成一个有机的城市结构

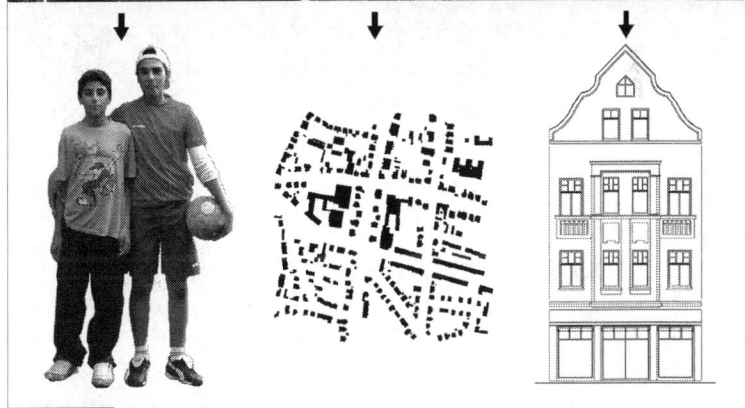

图3：
城市分析并不是对城市的整体复制

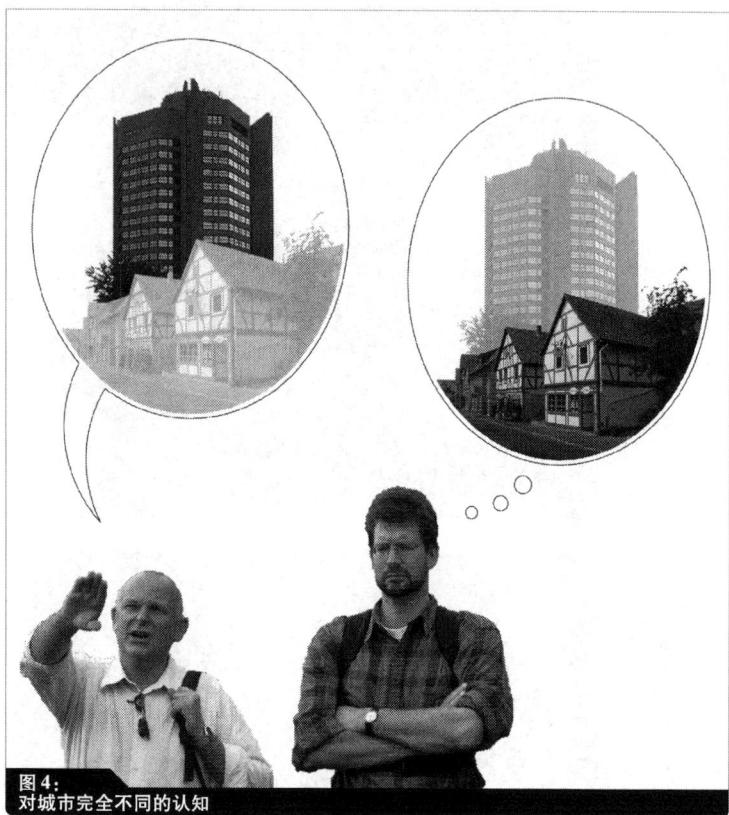

图4：
对城市完全不同的认知

　　城市分析可能会是城市正式规划过程的一部分。在进行公共项目开发、城市更新或者改造项目时，可能需要根据特定的需求进行城市分析工作。但是，总体而言，城市分析并不是一个标准化的过程，比如，城市分析的范围和深度必须根据每次的情况单独确定。

P13

城市分析的理论基础

　　一个城市能够用一个实体的可感知空间现象进行表达，而不深入分析和表达其形成原因；相反地，也可以将一个城市看成一些可用科学手段量测的因素发生作用的场所，而不考虑其空间尺寸。但在进行城市空间分析时，一般要避免以上两种极端的方式，而是将一个城市看作不同因素相互作用的产物，同时考虑其作用空间。一般来说，需要主要关注的是城市的可感知空间——因为离开了实体的可感知空间现象后，城市分析将无法进行。

美学观察：可视化城市

城市分析的一项主要任务是对城市的可视化感知和表达，比如，将视觉感知的城市结构进行可视化的记录和阐释。通常来说，我们在感知一个城市的时候并不需要了解感知机制，但在对城市进行表达的时候却与我们的感知行为相关。

从学术角度出发，感知是一个通信过程。本文中，感知指的是对视觉信息的理解。人们只能理解自己已经感知的现象，这点可以从感知过程的重复性特征中归纳得到，或者由观察者事先所了解或需要明确的。对于我们所认识的现象而言，其个体特征可能会随着时间的推移而改变，但一般特征必须是保持不变的。

我们的环境由信息的交流所构成。当我们听见一个事物的名称之后，会下意识地采用一种特定的方式来感知该事物所处的环境。同时，这种特定的方式也是该事物能够被感知的条件。从这个角度而言，人类的感觉器官并不是将其受到的所有刺激均传递到大脑中，而是类似于一个过滤器，对受到的所有感官刺激进行数据分类和压缩处理。

事物名称的确定并不是一个孤立的过程，而是以语言的形式并借助集体记忆记录在文化之中。同时，人类所具有的感知结构也建立在语言的基础之上。基于语言的分类，决定了我们在对周围环境感知、评价和阐明过程中的差异（选择的过程）和联系（构造的过程）。由于很多情况下这是一个下意识的过程，所以人们通常很难相信不同的人会因为经验和习惯的不同，而对相同的环境产生完全不同的感知。〉见图5

🖇

提示：
美学"Aesthetics"（来自于希腊语美学"aisthesis"一词，其意义即为感知"perception"）最初表达的是对美的学习。从科学角度出发，美学包含了人类所有感知的途径和机理。

图5：
人类的感知结构建立在语言的基础之上

　　许多我们经常用到的名词能够将城市所具有的差异非常大的组合到同一分类中。这与我们对人脸的感知类似，我们并不是要记住人脸中的每一个特征，而是记住了这些特征组合在一起时所留给我们的整体印象。采用类似的方法，当我们在感知一个城市的时候，我们会将其与许多我们已经熟悉的城市类型进行比较。打个比方，通常我们会设想有一个中心广场、古代教堂或者市政厅位于城市的中心。但实际上，这种情况只是一种理想化的城市形式，几乎在现实中无法看见，实际的城市中心通常由许多不同的城市现象所组成。而我们则按照自己的标准和熟悉的类型，选取了实际情况中与各自构想相符的若干部分，而与构想不符的则在感知的过程中被排除掉。打个比方，如果这些被排除的元素构成了我们的主要印象，那么我们在感知的过程中将不会把实际的城市中心当作城市中心来看待。〉见图6和图7

　　由此，我们可以得到以下结论：我们只能认知那些可以从已知事物中所抽象出来的现象。或者从另一种角度出发，我们只能感知那些我们所期望的现象。由于感知过程是一个文化现象，所以具有不同文化背景的人在进行城市分析时的感知可能会迥然不同。

　　在进行城市分析时，可能会通过强调某些现象的典型特点来对相似的感知类型进行分类。同时，通过了解控制感知过程的规则，能够帮助我们认识新的现象以及新旧现象之间的联系。

图 6：
我们在感知一个城市的时候，经常会将其与我们已知的城市类型相比较

图 7：
开姆尼茨的高速公路标志——城市的旅游形象一般是由现实中并不存在的一些元素组合而成的

科学观察：无形城市

城市中的大量个体现象相互作用就形成了我们对城市的美学感知。另外，科学方法只能对城市中的个体方面进行分析，而无法将整个城市作为一个整体进行分析。针对的范围越小，科学方法所分析的精度越高。科学工作者采用这种方法对处于感知结构之外或者无法被感知的现象进行分析。也就是说，感知过程是将所有的因素综合到一个整体印象中，而科学分析则是将整体的印象分解成若干单独的因素。打个比方，一个地区的特征可以采用其人口数量特征来表示，而不需要对该地区的居民进行切实的感知。〉见图8

定义值　　　科学的城市分析建立在对相关参数定义值的估计和阐释的基础之上，而相应取值的说明过程一般是建立在与其他调查方法所得结果或者其他时间对应取值的相互比较基础之上。分析统计学的运用则是一个非常典型的例子，比如对人口年龄或者就业率的统计分析。采用科学分析方法不能对城市空间中所有相关参数值的整体相互作用进行分析，但能够采用精确的量测手段和可供比较的数据来精确地对单独个体进行分析。

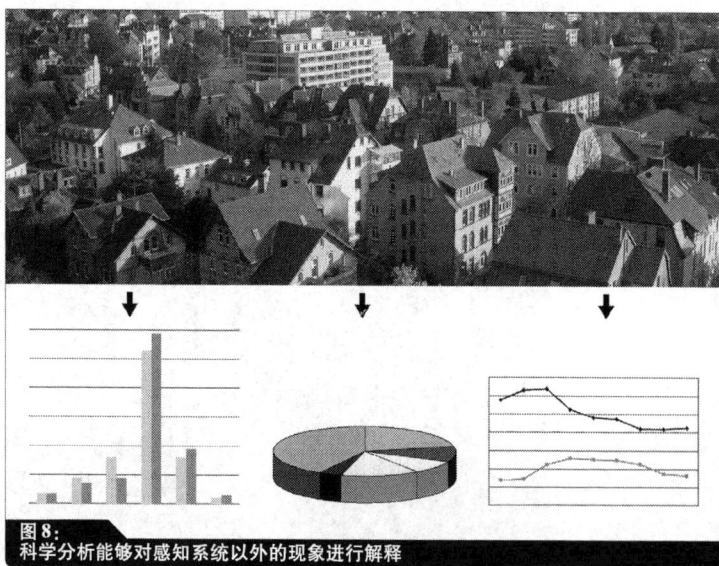

图8：
科学分析能够对感知系统以外的现象进行解释

整体观察：逻辑城市

形态描述

城市形态描述的是城市结构的空间特征，同时揭示这些特征形成的条件和原因，比如文化特征、政治特征以及地形特征等。因此，城市形态往往是以一种因果关系的形式出现的，即一些特殊的环境创造出一些特定的城市空间结构。

城市形态的分析需要对城市平面规划的相关历史进行大量的研究，比如街道的布置、房屋的平面位置等，而研究的内容则根据城市类型（有规划的城市、无规划的城市以及原生态城市）的不同存在很大的不同。其他相关的评价标准包括建筑的密度和分布情况、街道与广场的状况和等级、城市空地与建筑面积的比值，以及标志性建筑在城市规划中的位置等。一般，需要研究的主要城市结构特征包括：

　　—— **发展结构**：包括单体建筑、街区建筑以及沿街建筑等；
　　—— **通道结构**：包括街道、公共用地、广场以及桥梁等；
　　—— **空地结构**：包括空地和水体等。

虽然单体建筑的使用寿命相对较短，但整个城市的平面规划的变化却非常缓慢，对于一个街区而言，这种变化则更加缓慢。一个历史悠久的城市往往会具有很多不同时代的痕迹。由于城市街区规划一般随着城市的建立而开始，而不同时代具有不同的环境特点，所以现在的城市能够看到不同时代所保留下来的东西。这也同时说明，对于同一城市的形态描述，可能会得到不同的原因解释。〉见图9

提示：
　　形态学是关于形状和形式的科学。城市形态学所描述的是城市建筑和住宅的形式以及形成的过程。

提示：
　　在城市形态分析过程中，通过调查一个地区特征的形成条件和原因，能够推断出一个地区的发展过程。

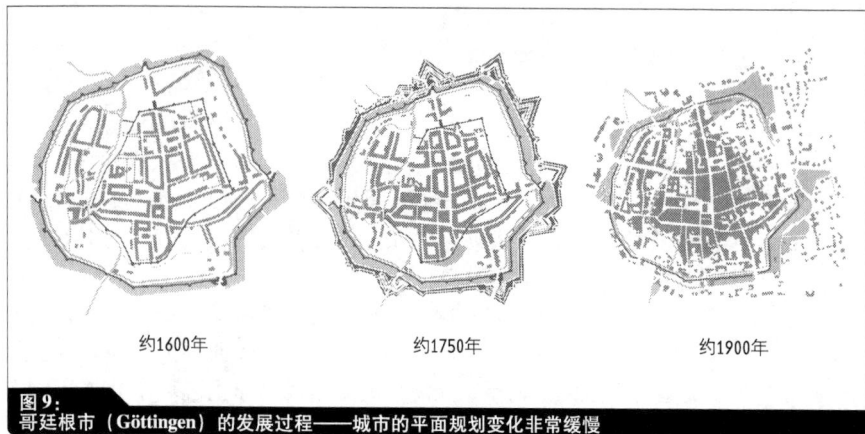

| 约1600年 | 约1750年 | 约1900年 |

图9:
哥廷根市（Göttingen）的发展过程——城市的平面规划变化非常缓慢

建筑风格分析

建筑风格分析是对不同时代（或者时代风格）、不同地区（或者地区风格）以及不同文化运动中的建筑或者城市结构所进行的形式上的综合比较。一般来说，城市的建筑风格不会由一个建筑师或者建筑部门的行政官员作为个体创造出来，通常是由文化运动中的少数派所创造出来的。所以，与其说建筑风格是一个城市空间主要的外形风格，不如说是一个城市传统建造经验在理性和形式方面的表达。

对于建筑风格的分析并不需要对每种风格的产生过程进行研究。虽然如此，从超越建筑风格的"装饰"层面含义的角度出发，一些特殊的世界观以及建造技术能够在阐释建筑风格时给予一定的帮助。

建筑风格的研究往往能够帮助我们推断建筑或者城市结构的建造年代。然而，并不是所有的时代都有各自特有的建筑风格。通常情况下，会出现多种建筑风格并存或者在不同的风格时代之间存在较长过渡期的情况。建筑风格和建造年代的确定在进行历史建筑的重新设计或者改建时会显得愈发困难，所以说建筑风格分析通常需要建立在深入的历史分析基础之上。〉见图10

图10：
建筑的风格能够帮助我们推断其建造年代

建筑功能类型分析

　　大多数的分析均认为城市的空间结构是其外部环境相互作用的结果，具体包括社会、气候、经济因素或者文化因素。而在进行建筑功能分析时则从两个方面同时进行，该分析认为：城市空间结构是社会和文化活动共同作用而产生的，同时城市的空间结构能够反映出社会和文化方面的行为。我们可以说，城市的空间结构是城市空间使用者进行社会活动的催化剂。

空间结构的可读性

　　人们并不是天生就明白社会活动行为应该与相应的城市空间结构相适应的道理，而是像学习语言一样，在一定的文化背景下学会了该道理。在改变空间结构的时候，需要保留其可读性，同时避免破坏位置与行为之间的关系。

　　城市建筑功能图示中最关键的一项是区分公共空间和私有空间，对空间结构的不同特征采用不同的标识方法，能够将其区分为公共空间或私有空间。比如，一个建有中心纪念碑、四周设有店铺的广场表明该广场为公共空间。〉见图11和图12

举例：
　　在进行建筑的初步设计时，同样需要考虑使用者的空间体验。通过家具、装饰物的陈设以及房间的位置，确保大家在首次使用时能够弄清楚不同房间的功能。相关规范对房间设计与使用者行为的联系进行了规定，在房间设计时不能忽视这些规定的要求。

93

图11：
即使设计存在不同……

图12：
……所有人都能明白建筑入口的功能类型

长期以来，特定的空间结构与社会行为方式之间存在相互联系。即便如此，仍然出现了许多新的建筑功能类型。这些新的建筑功能类型的出现一般均建立在对人类已经熟悉的建筑系统的基础之上。

由于建筑功能类型可能与特定的社会或者文化背景相联系，所以有些空间标识可能只有相应特定的社会或者文化团体才能读懂。因而，可能导致不同的团体成员对同一空间结构所对应的社会行为产生不同的理解，也正是这种理解的差异产生了不同的社会学说。想要弄清建筑的功能类型，需要对城市空间及其用途进行深入的研究。

P21 　　　空间体验与情感依恋

身体感觉和空间感觉既是一种身体体验，也是一种社会体验，同时它们也构成了所有人类体验的基础，人类所有的空间体验均来自这些基本体验。建筑与人类密切相关，建筑作为一个整体所反映的是人们相互之间的联系，而这种联系则反映在人们的体验之中，其中包含了各种正面或者负面的感受。

空间结构一般是在某种特定的情绪背景下被感知的，而空间格局则往往与生活方式相互联系。打个比方，我们会记住一个地方，可能是因为在该处经历了一件令人特别高兴的事情。又比如，我们会记住一个非常不起眼的地方，是因为我们在那里第一次遇见了未来的人生伴侣。

个体体验与
集体体验

以上这些情感联系往往是建立在个体体验的基础之上的,也就是说这种体验并不会在城市分析中起作用。但是,城市的空间格局同样能够激发出社会中的大量或者某些社会团体的集体情绪。在某些情况下,城市的空间格局甚至能够激起两个不同社会团体之间完全不同的情绪,进而引发冲突。

与所处的环境相比,这些地方往往显得格外引人注目。一般来说,这些地方要么是特别空旷的场所(比如平台),要么两种差异较大的空间结构的交界(比如河岸)。这种位置通常在空间中显得比较显著,给人带来一种突出的象征意义。在城市分析中,必须对这些能够引起强烈情绪的位置加以辨析和记录。〉见图13

图13:
能引起强烈情绪的地点标志

95

　　　　微观层面与宏观层面

对城市的感知可以分为不同层面。通常，人们对自己住所附近的区域非常了解，同时会将自己所了解的附近区域当作更大区域的一部分来看待，而并不会去对这个更大的区域进行详细了解。在对城市的一些结构元素（比如绿色廊道、河流以及交通主干道等）的感知时也存在相同的情况，我们会借助对显著点的了解，来对更大的区域进行感知。也就是说，城市空间中的所有部分均通过微观层面和宏观层面整合到了整个空间结构中。同时，不同的层面之间并不是相互独立的，城市空间中的任何地方都存在两个层面之间的重叠部分。

城市分析的实施

准备阶段与设计阶段

理论上说，分析对象中包含无数的信息量，故需要进行大量的分析工作。但实际上，我们在进行相关分析的时候，会在分析对象中建立对应的分析系统，而这些系统可以划分到历史、房屋地形以及社会结构等不同类别。>参考"分析方法"一章相关内容

主题确定

进行城市分析的主要目的包括以下两个方面：

— 揭示城市的整体状况。通过城市分析的方法，对整个城市或者某个区域进行调查研究，目的是得到该城市（区域）的整体状况。通常情况下，城市分析是针对城市空间中的错误所展开的。这些错误包括大量的城市空地、建筑维修不力以及城

> **提示：**
> 通过对房屋地形情况的调查，可以了解一个地区的空间结构和内部功能，其中比较重要的因素包括功能结构、空旷地区、绿地空间、已有建筑结构和发展结构等。

市人口的严重波动等。城市分析的目的是用肯定的判断方法来排除分析过程中的不确定，在这个过程中可能要对该区域使用若干不同的分析方法。

— 为某个特定问题寻找答案。即对特定的特征进行分析，比如已有住宅的状况、城市人口的社会状态或者拟建建筑的建筑规划的适应性问题。在这种情况下，城市分析必须对最初的问题提供一个明确的答案。

城市分析并不是一个标准化或者形式固定的过程，这就要求在进行分析之前，必须事先确定分析的主题。另外非常重要的一点是，需要根据城市分析的目的，选用正确的分析方法。如果不能确定适当的主题，可能会导致得到不恰当的结论，或者导致收集到的信息在分析过程中毫无用处。所以说在确定主题之前，需要对将要分析的区域有一个大致的了解。

P24
比例尺与界限的确定

城市分析所得的所有初始信息取决于分析对象的比例尺。打个比方，对于一个建筑小区分析而言，与单体建筑相关的信息可能有用，但对于整体城市的分析却没有任何用处。这并不是说小规模分析无法为所调查的区域提供正确的信息，而是说有效的信息必须能够在更广的城市背景中得以体现。较大区域中可能存在过多的信息——也就是所谓的"信息坟墓"——此时过多的信息反而扰乱我们所分析区域与其所在背景区域之间的关系。所以，在城市分析中非常重要的一部分工作就是将分析所得的信息进行压缩，找到重点信息。

整体城市分析中的最小空间单元是一个建筑小区或者局部地区（比例尺 1:20000～1:5000）。也就是说，我们所分析的所有城市特征都必须在该空间区域得以体现。如果我们将要调查的是某个局部地区，那么最小的分析单位通常是单体建筑或者建筑小区。如果调查的只是对象的少量特征，那么通常需要通过城市分析提供更加详细的信息。〉见图 14 和图 15

提示：
　　此处，"建筑小区"指的是被道路或者其他结构元素所包围的房屋社区。

图 14：
整体城市分析——整体分析的最小空间单元是
一个建筑小区

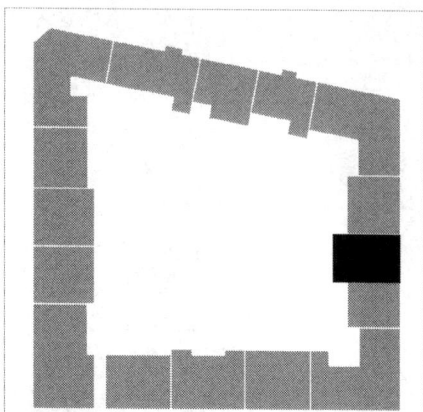

图 15：
建筑小区城市分析——分析的最小空间单元是
一栋建筑

　　所需信息的等级和详细程度取决于所调查对象的延伸范围。打个比方，采用的比例尺越大，所调查的区域范围越小。然而，一个城市（或者更广的范围内）是一个整体，所有的规划区域都不是自我独立的。所以说，在进行城市分析的时候需要考虑到考察对象以外的区域，并需要考虑其与整个城市结构的相互关系。〉见图 16 至图 18

图 16：
在对调查对象进行分析的时候，必须将邻近区
域列入考虑的范畴内

图17：
同一地点活动着与不同城市区域相关的城市现象

图18：
住宅、街道、城市以及流动的水道

P25   延续与持续监测

　　城市分析的延续包括以特定的时间间隔对建筑进行调查分析。延续分析可能会采用之前相同的分析方法，或者是原有方法的补充或改进。方法的不同可能会导致分析重点的改变。

　　城市分析中的监测指的是对调查区域进行的长期观察，通常采用的方法是对一些数据以固定的时间间隔进行持续的量测，然后对所量测的结果进行比较，并与其他地区的相关结果进行对比，或者是将量测结果相同的若干调查区域进行对比。典型的调查数据包括居民数量、人口统计数据特征以及社会人口数据特征等。

提示：
　　调查对象区域的空间整体情况可能会对该区域产生积极的影响，比如该地区附近所存在的娱乐设施等。但有时候也可能会带来冲突，比如在住宅区附近设有排放烟气的工厂。

提示：
　　为了能够将监测结果进行比较，在监测过程中必须建立明确的评估参数和比较方法。因此，在监测结果比较时，一般仅仅采用标准化的参数值；同时还需要确保不同调查区域之间存在相互的可比性。

设计专家之间的合作

通常情况下，由于城市分析涉及许多不同的专业领域，一个设计专家很难单独完成一项城市分析。也就是说，在进行城市分析准备工作和日程安排的时候，必须确定出所需的设计专家的准确数量。在实际操作时，通常采用的是建立工作团队或者将工作划分为与市政规划部门相关以及与设计专家相关两部分的方式。一般来说，城市规划专家、建筑师和房屋地理位置设计师将分别监督城市分析的几个主要阶段——土地用途调查、交通情况调查、房屋结构形式以及地理位置确定等。与此同时，可能需要其他专业分析人员，比如交通系统规划专家或者历史学家等。与城市空地或者生态相关的问题一般需要相应的城市空地规划师或景观设计师来解决，同时还需要生物学家、气象学家以及地理学家在其中担任部分工作。如果需要进行城市社会结构分析，则需要社会学家来担任大量的调查和统计工作。当然，在工作时需要对不同参与者之间的主次关系进行调整。

数据的获取与处理

城市分析的过程中需要搜集大量的数据，所以应该尽早取得获取相关材料的许可，或者尽早做出安排，以便相关部门能够在其接纳能力范围内批准相关人员进行数据搜集。

在城市分析过程中需要搜集和处理的数据形式种类繁多，比如：

— 纸面文件数据，比如打印版的报告、设计图等；

— 电子数据，比如文档、图片等相关设计文件；

— 亲临现场获取的相关信息（参考"在分析对象地区进行的工作"一节相关内容）；

— 与第三方谈话获取的相关信息（参考"在分析对象地区进行的工作"一节相关内容）。

注释：

在建立工作小组时，需要对研究领域进行明确划分，以免出现人员冗余。由于某个方面专家的工作成果可能会对其他人的工作产生一定的影响，在必要的时候可以建立严格的工作时间表。

注释：

在城市分析工作开始之前，应该弄清楚提供或者准备所需数据材料的管理部门以及主管人员。

在将城市分析的结果进行公开发布之前，需要检查分析过程中所使用材料的版权问题，必要时应该在结果文件中加以明确说明。私人相关数据受到相关数据保护法律的保护，如果涉及这些数据，必须与相关法律保护部门取得联系，并在得到相关许可之后才可以使用。

P28   历史信息

一个城市的历史地图、出版物以及编年史是进行历史分析时非常重要的参考资料。通常来说，比较好的方法是到城市档案馆或者图书馆查阅相关资料。在一些小城市，这些工作通常由一些名义上的任命人来进行，然而他们一般只能提供照片、当地出版物等材料，而无法提供一些处理过的科学信息。由于对这些原始历史材料的处理和评估是一项非常繁重的工作，所以一般只有在需要进行科学调查分析的时候才进行这项工作。

P28   分析的基础——地图

比例图是城市分析中一项非常重要和实用的工具。地图本身不仅能够提供信息，而且还能够作为背景将不同的分析结果融入其中。无论国内外，均没有相关标准规定什么样的地图可以用于城市分析，所以说在进行城市分析之前，我们应该对所采用地图的实用性和技术质量进行检查。由于不同地图所采用的投影制图法可能不全相同，所以并不是所有的地区之间都能够相互兼容。也就是说，在一项城市分析中不能一直使用同一张地图，而是根据研究的实体分别进行地图的绘制和选用。

地形图   地形图主要表现地球表面的物体和地理特征。由于图中具有等高线，使得地形图在城市分析中起到了非常重要的作用。在城市分析中一般用到的地形图比例尺在 1∶5000 到 1∶50000 之间，而地形图一般采用的数据网格形式，有时候可以将地面物体或者地理特征进行分层管理，在进行打印时则各自对应一组数据。〉见图 19

提示：

如果调查对象区域覆盖了具有重要文化意义的物体（建筑或者地点）或者是已经登记的大地水平面等，那么相关政府部门一般有义务提供相关的详细信息和记录。

提示：

由于地形图的表达方式通俗易懂，所以在一些公共工程中可以不用改变地形图的表达形式而直接采用。

图19：
地形图

图20：
房屋图

房屋图

地理数据的记录是以房屋图作为基础，并便于与其他专业数据进行相互联系和交换。房屋图是土地登记信息的一种可视化表达方式，表明了每一块土地的所有权情况。

房屋图中表达了房屋的空间、地形特征以及土地的划分情况。由于房屋图中包含了地面物体以及土地小区的大量信息，故在进行城市分析的时候往往会非常愿意采用房屋图。地形图中重点标注了土地和地面物体的性质和特征，比如，土地的用途、建筑的层数等，并将属性相同或者相似的地面物体归为同一分类中。然后根据物体的不同分类进行房屋图的分层处理，同时也将图中的内容按照主题进行相应分类。房屋图同时也是土地登记图形信息系统中的一个组成部分，打个比方，房屋图中除了其可视化的信息之外，还包含大量的数据信息。〉见图20

房屋图一般是以电子版的形式由土地登记部门保存，将其转化为矢量文件形式之后可以由其他使用者进行编辑。在参考房屋图之前，我们应该提前弄清楚，从中能够得到哪些与地面物体或者小区相关的信息。否则在参考房屋图的过程中，还需要对大量的数据信息进行排除。

一般，CAD（计算机辅助设计）系统中常用的文件格式为 ＊.DXF（图形交换格式）与 ＊.DWG（图形格式，Autodesk 公司系列产品的一种文件格式）。如果需要对房屋登记部门的信息系统文件进行格式转

102

化，则需要涉及其他种类的数据格式。

航拍照片　　　　　航拍照片可以采用垂直或者以某一倾斜角度的方式进行拍摄。因为具有便于与其他图形资料结合使用的优势，更常用的是垂直拍摄纠偏的真实比例照片——所谓的"正射影像"。结合房屋图，航拍照片能够提供非常重要的信息——地面物体的坐标，特别是能够准确地指出绿色基础设施的地面位置。另外，对于非专业人士而言，采用航拍照片可以更加有效地解释城市分析的结果。〉见图21

详图　　　　　根据调查区域的不同，有些情况下城市分析可能会用到大比例尺的详图。比如，市政工程需要提供详图来标明地下管线的布置情况。

地理信息系统　　　　　地理信息系统（GIS）用来管理可视化的空间数据信息。基于系统中的信息，可以进行大量的评估工作。地理信息系统的首要工作是采用数据库的形式对大量的数据信息进行存储，并通过评估的方法将不同的数据之间建立联系，而采用地图形式表达信息并不是地理信息系统的主要工作。不同数据之间主要根据各自的空间特征建立相互之间的联系，数据的空间特征则通常以几何图形的形式得以存储——比如采用 CAD 系统，而这些几何图形通常是在标准图（比如地形图）的基础上绘制而成的。

　　　　　利用 GIS 系统进行城市分析需要完成大量的前期研究工作，另一项非常重要的工作是确认数据之间的兼容性。否则，在进行数据格式转换的时候可能会导致数据的丢失，或者需要大量的人工数据导入工作。公共实体能够提供大量的公用数据信息，但私有建筑所提供的空间数据信息日益增多。个体所进行的调查结果能够合并到 GIS 系统以及其他的外部数据库中。〉见图22

提示：
　　如果房屋图所涉及的面积较大，在进行最终数据提交之前首先进行测试数据的交换（压缩）可能是一个比较好的做法。

提示：
　　对于非专业人士而言，房屋图的专业性质较强。所以在采用房屋图进行分析结果说明时，需要进行大量的修改工作，来将房屋图进行图形化处理。

图 21：
垂直拍摄的航拍照片（正射影像）

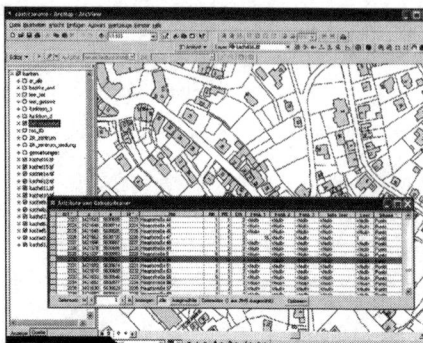
图 22：
GIS 系统中将空间特征存储在数据库中

现场计算机　　　　　采用现场计算机，能够在调查现场对标准化调查结果进行数据处理和分析。输入计算机的信息可以通过点击触摸屏所示地图或者内置 GPS（全球定位系统）接收器计算匹配的方式，赋予到正确的位置中。然后，由 GIS 系统对输入的信息进行评估处理。〉见图 23

P31　　　　　　　　已有规划设计与分析

通常情况下，拟将调查的区域在此之前存在相应的规划设计与分析结果，而这些材料必须根据相关规定和信息进行评估。正式的规划工作必须在遵守相关规定的基础之上进行，比如，建设用地的使用范围和性质、周围环境需要保留的状态特征以及存在洪水风险的地区范围等。有些情况下，还存在相应专家对该地区的一些重要参数和周边环境的评估结果，这些结果主要包括：

提示：
　　采用 GIS 系统对一个地区进行长期的观察（监测）是一项非常有趣的工作，因为 GIS 系统中包括最新数据，非常便于对该地区进行新的评估以及生成新的评估图。

图 23：
一种现场计算机

— 生态专家对已有重要绿色基础设施的评估文章 (参考"城市空地
和绿色结构分析"一节相关内容)；

— 针对噪声污染所进行的噪声防护报告——已有报告或者在规
划工程之前预先完成；

— 已有交通压力与交通设施、交通网络和交通系统情况下的交
通状况分析报告 (参考"交通结构分析"一节相关内容)；

— 调查区域零售商业结构的分析报告 (参考"土地使用结构分析"一
节相关内容)。

以上已有评估报告中的相关信息都将合并到城市分析的过程中，
并将对这些信息与其他因素以及将来城市规划之间的关系进行分析。
根据重要性的不同，部分信息将在比例图中得以体现。

**人口数据与
统计**

通常情况下，行政部门应该对人口结构进行常规性的统计分析。
统计分析内容包括地区的整体人口情况以及某个主题下的专项统计分
析。一般情况下，进行城市分析的区域与进行人口统计分析的区域不
会完全一致，这也限制了人口统计分析数据在该城市分析区域中的应

提示：
　　实际上，在进行城市分析的最初阶段就应
该向有关部门申请获取已有的城市规划和分析
结果。另外，大量的城市规划内容能够从因特
网上下载得到。) 见图24

105

图24：
对于城市分析而言，正式规划文件中具有大量重要的信息

用。为了对城市分析区域的相关人口数据进行评估，必须重新完成一项人口普查工作。

但人口普查的区域必须由行政管理部门来进行划分，因为该部门之外的专家没有访问居民数据信息的权限。为了满足相关数据保护要求，只有在获得了城市分析专家和相关权力部门授权之后，才有可能获取足够的数据信息，完成小范围的调查分析。

进行人口结构分析非常有效的一种方法，是将特定调查与调查结果随时间变化规律的研究相结合，并将调查结果与其他地区的相关结果进行对比，以确保能够获得最有效的对比数据。

提示：
"人口普查区域"值是一个能够赋予诸如人口平均年龄、家庭数量等特殊属性的连续区域。

　　　　在分析对象地区进行的工作

编制现场工作清单

现场调查工作是城市分析中一项非常重要的工作，而且大多数分析结果只能通过对现场情况的调查分析才能获得。比如，某处的空间印象和大气质量只有当分析人员身临其境时才能感受到，所以有时候可能需要在一天的不同时刻或者一周内若干天对某处进行调查。

在进行现场调查以前，需要进行相应的文件准备工作，以用来记录调查所得的文字和图形信息。根据城市分析的范围和深度不同，需要记录的信息量和空间性质也随之不同。所以说，我们所准备的文件应该与涉及的空间关系和信息量相适应。如果准备的文件包含过多的信息，可能会将随后的评估工作转变成非常费力的信息侦查工作。〉见图25

一般来说，进行现场调查的分析人员会对自己所关注的内容进行评估。但对于范围更加广泛的调查工作，可能需要组成一个调查小组。对于这种情况，需要事先统一分析结果的记录方式，以防止不同人员之间对分析结果产生不同的理解，而标准化问卷则是一种非常适用的调查方式。

现场调查分析的结果可以记录在地形图或者房屋图上，根据分析范围以及深度的不同，可以采用不同比例尺的图形：

—— 对于城市整体分析，一般可以采用比例尺在 1:5000 ~ 1:20000 之间的地形图或者航拍照片；

—— 为了表示房屋的地理特征，一般可以采用比例尺 1:1000 的房屋图。在进行图形准备的时候，可以将与现场调查无关的图层删掉；

注释：

在进行现场调查时，为了防止发生信息遗漏，在调查之前应该进行充分的准备工作。否则一个小的失误就可能导致长期的额外工作，特别是当调查地区路途较远的时候。

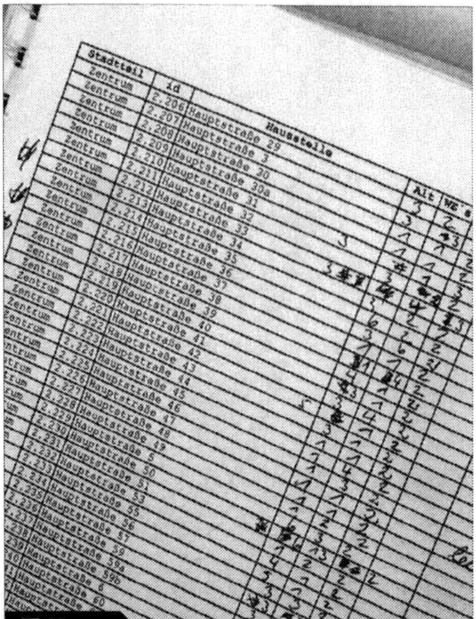

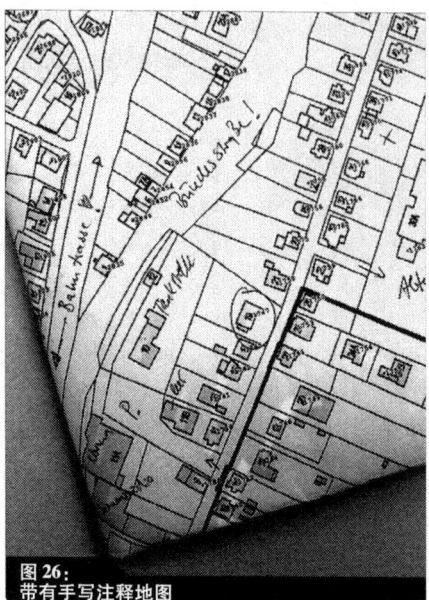
图 25：
房屋特征调查记录簿

图 26：
带有手写注释地图

— 对于记录建筑相关数据的房屋图，一般采用 1:500 的比例尺。
如果采用更大比例尺的纸版地图，则会给现场工作带来不便。
如果允许在地图上进行小的装订，同时也允许将分析结果列
在调查表中，那么将地图和调查表装订成整体可能是一个比
较好的方法。调查表一般会采用标准化问卷的形式，以避免
在编制材料清单的时候发生遗漏。〉见图 26

相机一般用来标明所调查的地区。一个比较好的方法是在地图上
标明照片拍摄的地点，便于在随后的分析中，将照片插入分析文字的
正确位置。

P36

居民参与以及与当地管理者的联系

当地居民对自己所居住地区的邻里环境有着非常详细的了解。另
外，该地区的商人、地区自治管理人员也对当地条件以及对人们的影
响有一定的了解。在所有的城市分析中，都应该包括以上信息。

可以采用不同的方法与当地居民取得联系，这种方式在我们评估
调查地区的各种因素和冲突对人们的影响的时候，比现场信息搜集的

帮助更大。

居民会谈　　居民会谈是市民参与的一种常见方式，而且不需要太多的准备工作。但实际上，如果没有针对特殊的主题或者规划活动，参与的市民将不会产生太深的感受，所以参与的积极性和效果也不会太好。所以，在组织居民会谈的时候，建议对具体的测量工作以及大体的规划工作进行讨论。另外一种建议的做法是将初步的城市分析报告展示到参与市民的眼前，为会谈搭设一个讨论的框架。在有些情况下，为了避免将分析结果曝光，采用这种方式的时候将不会为参与市民提供分析结果的样品。在会谈地址的选择时通常不需要太高的限制级别——一般可以选择在调查区域展开。

个人公共咨询　　个人公共咨询的准备、实施和结果评估需要消耗大量的时间，所以只有在需要进行大量社会结构分析的时候，才会进行个人公共咨询工作。此项工作的进行和结果评估需要得到社会学家的协助。

　　个人咨询可以采用访谈或者标准问卷调查的形式，其中标准问卷调查的结构评估相对更加容易一些。但若采用大量邮寄的问卷，则会导致调查结果的返回较慢，同时调查对象的抽样也没有太强的代表性。

专家咨询　　采用专家咨询的方式，能够在短时间内将广泛领域的大量专业知识进行汇总。专家咨询一般是邀请相应领域的专家对一些特殊的主题进行集中讨论。所邀请的专家一般包括讨论领域的专家以及活跃分子——这些人可能来自社会福利组织、志愿工作者或者是当地企业家。

注释:

　　在进行区域现场调查的时候，可能会与当地居民产生一些自发的接触。我们不应该回避这样的见面，而是进行仔细的咨询，有时反而会得到一些非常有价值的信息。

分析方法

对于某一地区进行城市分析可能包括一个或者多个调查主题，比如调查地区不同的土地用途或者交通组织。前文已经提到，这些分项分析都是对现实情况的一个简化抽象和局部代表。实际上，由于无法考虑所有因素之间的相互影响，这些分析过程都进行了简化。同时，在进行城市分析的时候，需要将不同分项分析之间的相互影响尽可能考虑在其中。简单来说，城市分析在最初阶段就像拆解汽车一样将调查区域进行分解，然后根据对每个部件研究所得到的信息加上不同部件之间的相互关系，再将不同部件组装成一辆汽车。

由于城市分析并没有固定的方式，所以实施起来有无数种不同的方式。因此，下文将只涉及实际市镇规划中最常用的分项分析。

历史分析

历史分析主要对调查区域产生、融合的历史成因或者具有重大影响的历史事件进行分析。同时，近代历史上发生的事件所造成的影响也可能会比较大。历史分析不仅仅局限在现有城市风格成因的分析上，而且还将记录分析对调查区域空间发展具有潜在影响的事件。〉见图 27 和图 28

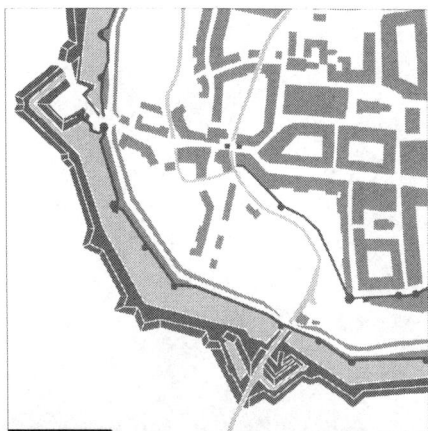

图 27：
哥廷根市过去的城市防御工事……

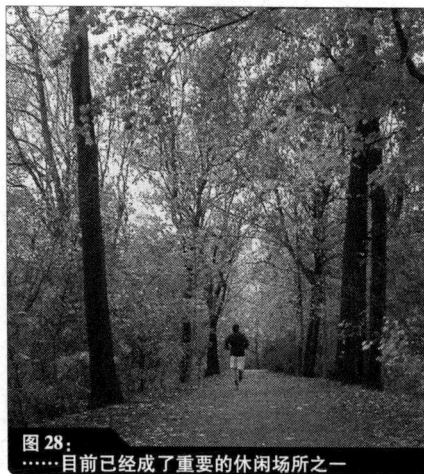

图 28：
……目前已经成了重要的休闲场所之一

以下是进行历史分析时可以采用的两种可能方式：

— 将调查区域作为整体进行分析：这种方式需要对调查区域整体结构造成影响的历史事件和环境进行分析；

— 将调查区域作为不同部分的组合进行分析：这种方式需要对城市结构的单独特征进行分析，确定其是否对城市发展产生过（或正在产生）典型的影响作用。

历史事件包括对城市空间所采取的一些措施，比如修建城市防御工事、城市火灾后的重建等。然而，许多历史事件并不是城市空间中所采用的特殊措施，但也对城市的空间发展产生了一定的影响。比如，整个地区的经济繁荣或者萧条都可能会对该地区的空间发展造成一定的影响。

个体特征的记录

一般来说，具体对调查区域采用整体分析还是各点单独分析，主要取决于该地区所拥有的重要历史建筑的数量。在点对点的分析中，对不同建筑进行单独的评估，并记录所有与历史相关的特征。而对建筑的评估一般是遵循一定规则的标准化过程，然后——如果有必要——将建筑划分到不同的建筑类型中。通常采用的规则包括建筑的年代、建筑历史风格的保留程度和状态等。接下来，需要在地图中对应的位置，根据所具有的特征对建筑进行标注。这种方式可以协助确定不同建筑类型在实际空间中的分布情况，以及不同重要建筑组的数量。〉见图 29 和图 30

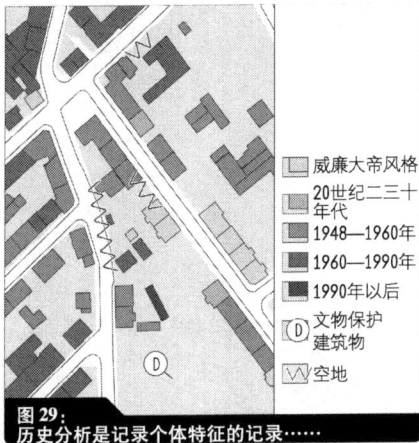

威廉大帝风格
20世纪二三十年代
1948—1960年
1960—1990年
1990年以后
Ⓓ 文物保护建筑物
空地

图 29：
历史分析是记录个体特征的记录……

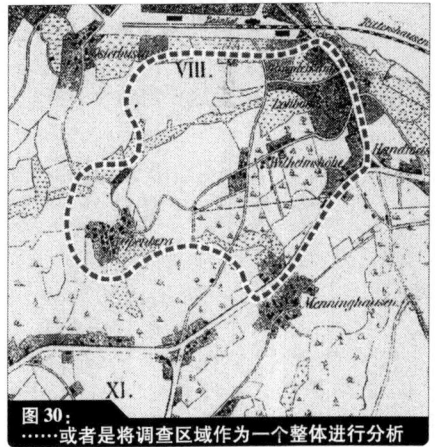

图 30：
……或者是将调查区域作为一个整体进行分析

111

在极个别的情况下，某个调查区域的空间特征能够解释成某个历史事件的结果。比如，在一个有限的时期内，将全面规划的地区修建完毕。但是，对于调查区域的空间特征更多的是在若干不同历史事件和环境影响下形成的。而这些历史事件通常会发生在不同的时期，从而对调查区域进行了一次次的改造，同时在调查区域中的不同地点（不同时代的建筑）留下了痕迹。另外，当出现了事件搭接和重叠时，不同事件也可能对同一地点造成影响（比如20世纪50年代采用中世纪建筑平面设计所建造的建筑）。根据分析得到的个体特征的分布区域和分布密度，能够对调查区域的发展是否能够追溯到较小数量的历史发展以及历史事件上，或者说现有的城市特征背后是否具有不同的历史发展类型。

历史分析还包括对调查区域与周围地区或者整个城市之间相互关系的研究，其目的是明确该城市区域发展的背景情况。但在此之前，必须确认调查区域在整个城市的发展过程中具有的典型特征，或者是扮演的特殊角色。

对一个城市的规划历史研究而言，非常重要的一点是对调查区域进行充分的历史研究。但是，在进行历史分析、论证调查区域的特征以及未来发展方向的过程中，也可能会发现，有些最初被认为历史意义不大的区域能够揭示该城市的典型发展过程。在进行历史分析结果解释时，非常重要的一点是对已经对调查区域发展产生了明显影响的事实进行说明。而那些对于城市历史的笼统讨论，比如朝代更迭、参考文献或者城市结盟信息等对于调查区域的城市性质基本没有影响，所以应该排除在城市分析的范围之外。

土地使用结构分析

土地使用结构分析的内容包括对调查区域的土地真实使用情况、主要使用情况以及不同使用性质之间关系的研究。

🖉
提示：

在对调查区域作为一个整体进行分析的时候，一般会采用历史地图来对城市的历史发展进行分析。由于比例尺的不同以及复制技术的限制，一些地图一般无法与现代地图相提并论。

🖉
提示：

土地使用强度一般与建筑密度有关，这意味着应该将建筑和小区的分析结果与城市使用结构的分析结果进行对比。

土地使用
强度

土地使用强度可以采用统计分析数值进行描述，具体可以采用调查区域的统计处理数据而生成的比例值。其中一种方法是计算每个具有不同使用性质的土地所占调查区域总地表面积的比值。另外，土地使用特征的统计可以采用如下方法——比如，工作场所的数量等。土地使用强度还可以用密度值的方法来进行描述，具体是计算单位地表面积上具有某种特征的土地的关系值，比如单位土地面积内的人口密度是最常用的一种密度值。

个体特征的
记录

土地使用结构分析还包括确定空间单元的实际用途。此处提到的"空间单元"可以是一块空地，也可以是一栋建筑。如果某一块土地被一栋建筑所占据，那么决定该土地使用性质的主要因素将是这栋建筑的功能。打个比方，如果某处建筑含有公寓住宅，那么该区域土地将会被划分为"住宅用地"，无论该建筑（小区）附近是否存在空地。

对于整个城市或者具有一定面积的调查区域（比如，包含若干建筑小区的区域）所进行的分析中，包括对建筑小区（或者建筑邻里）的土地使用结构分析。区域土地使用结构分析通常是以建筑小区为单位逐个进行的，比如：确定每个建筑小区的土地用途，或者确定具有某一特殊土地使用性质的建筑小区数量等。在分析过程中，需要对交通系统所占地表面积、绿地面积、建筑小区面积等进行统计。

土地使用结构分析可能只对建筑使用性质进行分析，而将空地的使用性质分析放在空地和绿地结构的分析中。这种分工方法在采用分层方法记录土地使用功能的时候非常有效，而这种详细程度通常只用在调查区域面积很小的情况下。〉见图31和图32

图31：
在进行土地使用结构分析的时候，根据比例尺和详细程度的不同，分析过程可以针对若干建筑小区逐个进行……

住宅用地
多种用途
商业用地
附属设施

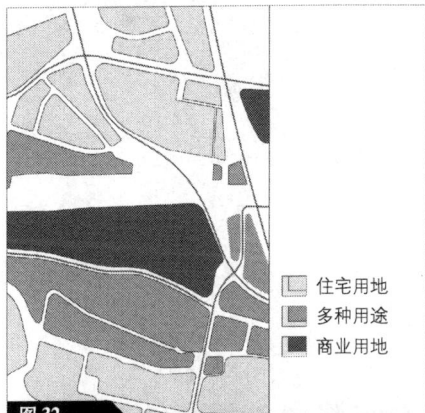

图32：
……也可以针对某一个建筑小区进行

住宅用地
多种用途
商业用地

113

土地的使用性质一般可以通过现场调查获得，比如通过眼睛看。但是，从房屋图数据中能够得到更多的信息。在有些情况下，相关政府部门可以利用 GIS 系统提供一些地面使用性质的相关信息（参考"数据的获取与处理"一节相关内容）。另外一个信息来源是某些商业机构的公司产品目录，但在使用的时候存在一定的风险，所以需要确认产品目录中的产品邮政地址在调查区域范围之内，而不是产品的生产地址。

可以根据所需的详细程度对调查区域进行不同程度的土地使用结构分析结果展示，具体操作时建议根据主标题和副标题将分析结果进行划分。〉见图33

住宅用地

一般情况下，住宅用地的使用可以根据住宅的建筑形式进行分类，比如：独立式住宅、半独立式住宅、联排住宅以及开放式或者封闭式的多层公寓等。对于<u>特殊的住宅形式</u>——比如掩蔽体或者住

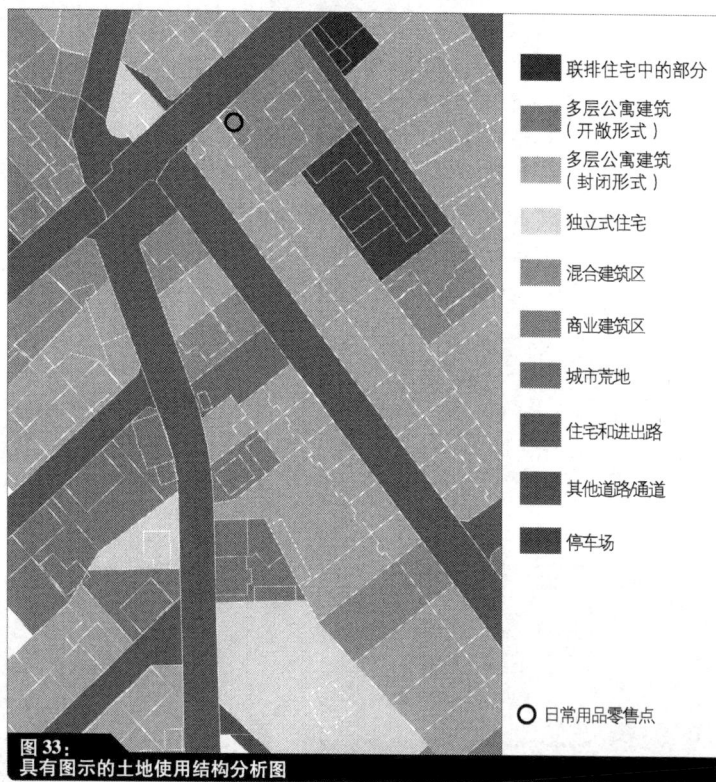

图例：
- 联排住宅中的部分
- 多层公寓建筑（开敞形式）
- 多层公寓建筑（封闭形式）
- 独立式住宅
- 混合建筑区
- 商业建筑区
- 城市荒地
- 住宅和进出路
- 其他道路/通道
- 停车场

○ 日常用品零售点

图33：
具有图示的土地使用结构分析图

所——一般都需要进行单独标记，因为这些住宅对当地的供应商以及交通系统存在一些特殊的要求。

商业用地　　　可以根据对土地使用造成的干扰程度来对商业用地进行划分——比如，根据与住宅用地之间的协调性程度等：

— 工业贸易区对土地造成的扰乱程度最大，所以往往将工业贸易区按照规划要求严格控制在工业区。

— 贸易区则包括各种具有不同商业用途的土地。这些区域一般位于工业区，土地使用性质主要包括制造、运输、仓储、批发、城市补给以及手工业等。

— 非破坏性贸易区包括与住宅用地相配套的各种商业用地。绝大多数的使用性质是服务或者零售。

服务区用地　　　服务行业的产品具有非物质性，主要取决于人与人之间的联系。因此，服务行业能够根据其与社会公众之间的联系程度进行分类：

— 面向顾客的服务行业指的是那些需要经常与公众打交道的行业，比如旅游代理或者休闲娱乐等；

— 非面向顾客的服务行业主要指的是办公。

服务区的使用者也可能是供应商，所以服务行业也可以划分为供应服务和非供应服务。具有供应功能的服务行业包括医疗诊所、邮局以及包裹物流公司等。

零售点　　　零售指的是向最终用户出售产品的商业行为，而批发则指的是向销售商出售产品。

对于小区层面的分析，必须对出售日用生活必需品的零售点布局进行分析，因为零售店是该地区居民重要的生活福利设施。对于小区内所有位置的居民而言，日用生活必需品销售点距离均不应该超过500m（良好生活福利设施布置的标准）或者700m（合格生活福利设施布置的标准）。日用生活必需品包括食品杂货、日用奢侈品、饮料、医疗以及清洁产品等。而医疗诊所、邮局等设施也有可能划分到当地生活福利设施的范围之内。

> **提示：**
> 零售业按照不同的标准进行多种不同方式的再次划分，比如根据市场的形式（自助、超市、专营商店、专营市场等）、中心位置、服务区域、分店情况以及商店面积等进行划分。

混合区	混合区指的是住宅用地和其他用地（一般是住宅配套商业）组合使用的土地。根据城市分析详细程度的不同，还可以按照不同的组合方式将混合区进行进一步划分——比如，住宅与旅店用地、住宅与办公室用地等。通常，还会对混合区内的组合情况进行分析，以确定该混合区是以住宅还是商业为主。
公共福利设施	公共福利设施指的是面向大众开放的公共设施——最常见的包括幼儿园、托儿所、教堂、文化设施和民政机关等。另外，一些户外设施，比如游乐场地、墓地等也有可能划分到公共福利设施之内。 休闲和社交场所既可能是公共福利设施，也可能是由商业机构运营。根据其性质的不同——公共或者私有——可以划分为公共福利设施或者面向顾客的商业服务设施。
旅店与餐饮设施	餐饮设施包括所有能够提供餐饮服务的设施，比如餐馆、小吃店、咖啡厅、酒店以及冰淇淋店等。旅店指的是能够提供客房以及相关服务的设施。
娱乐设施	娱乐设施指的是用作娱乐和放松的场所，比如歌舞厅、游乐场等。由于这些场所将对分析和规划造成较大的破坏性，所以在进行城市分析的时候通常将娱乐设施作为一个单独分类进行对待。
设备设施	设备设施指的是用来调节和维修城市区域不同供应设备或者控制特殊商业过程的设备或者建筑，包括燃气调度站、变电站、污水处理厂等。大多数的设备设施外围需要设置保护区域，比如设备设施与周围其他性质的用地之间必须满足一定的距离。
交通设施	交通结构分析将会对不同的道路类型或者道路所具有的不同财政性质进行分类。不论哪种分类方法，城市土地使用结构分析的结果都能够为交通设施的设置提供参考。 与交通相关的建筑，比如火车站、公交汽车总站等可能会划分为交通设施，也有可能划分为公共福利设施。
城市空地使用	一般来说，城市空地使用包含在城市空地和绿色结构分析过程中，但也可以放在土地使用结构分析过程中进行。城市空地的使用包括停车场、游乐场所等，而这些场所可以划分到公共福利设施之内。
闲置土地	闲置指的是功能的缺乏。若一个地区存在大量的闲置土地，则说明该地区缺乏吸引力，而导致的原因往往是规划的破坏性或者规划上的失误。但一般很难判断和证明土地的闲置，一部分原因是因为用眼睛很难轻易地判断出是否闲置，另一部分原因是为了确定一栋建筑的

闲置时间需要进行长期的观察。

住宅用地的闲置与商业用地的闲置性质不同，因为它们通常由不同的原因所造成，相应需要不同的措施来改善。另外一种比较明智的做法是对一栋楼的部分闲置和整楼闲置进行区分。〉见图34

其他用途土地

对于调查区域中仅仅出现一次或者少次的其他用途土地，一般采用在评估地图中直接标出的方式，以避免造成地图中图示混乱或者长度过长。

车库、掩蔽处以及其他类似的设施可以划分为附属设施。

颗粒化分析

土地使用结构分析的主要任务是对类似的土地使用类型进行细分，并对土地进行颗粒化分析。其中，非常重要的一项工作是标明调查区域中的主要地区，以便于区分不同使用性质土地的破坏性。不同的使用性质会给土地带来不同的破坏性等级，但同时也给土地带来了相应的受保护权利。当使用梯度较大时——相邻土地的破坏性和受保护权利差异较大时——通常会导致冲突；而当使用梯度比较平缓时——相邻土地的破坏性和受保护权利相似时——则导致冲突的可能性较小。

调查区域的背景分析

在对类似的土地使用类型进行细分时，需要将调查区域相邻的地区也纳入考虑的范围内，因为这些相邻地区也会带来相应的破坏性和保护权利。

一个地区的颗粒化分析反映的是该地区的土地用途分布以及该地区的土地使用平衡关系，同时还能够在更广范围的背景内表达出该地区的土地使用目的和地区的中心位置。根据规划类型的不同，可以将调查地区进行如下分类：〉见图35

> **提示：**
> 除采用观察的方式之外，还能够通过监测建筑的电力使用情况来确定土地的闲置情况。但这种方式需要供电单位的参与，所以通常很少采用。

图 34：
房屋的闲置说明该建筑可能无法满足建造意图

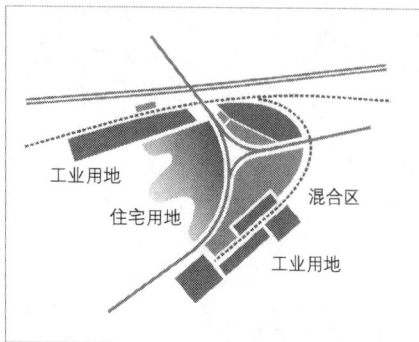

图 35：
区域整体状况的抽象性表达

工业用地
住宅用地
混合区
工业用地

— 单纯住宅用地；
— 设有少量配套设施的住宅用地；
— 混合区；
— 内城区与市中心；
— 商业区；
— 工业区；
— 特殊区域。

总结　　　　　在进行最后评估的时候，必须明确进行区域调查的目的，并由此判断调查结果中哪些是支持我们的调查目的，而哪些是无关的或者相悖的（参考"分析结果的解释与表达"一章相关内容）。进行土地使用结构分析的中心工作之一是找出所有的土地使用冲突，但在此过程中必须分清导致冲突的原因（比如，相互之间的不协调）和冲突的后果（比如，增加了土地的闲置）。〉见图 36

P48　　　　　**交通结构分析**

交通结构分析的主要工作内容是对调查区域的已有交通设施进行分析，以及对调查区域与交通网络之间的联系和整体性进行分析。从这个角度出发，道路既属于交通设施，也属于公共空间，所以道路在调查区域的感知过程中对道路的感知尤其重要。因此，城市分析的目标中也包括了对道路的设计。通常情况下，每条道路具有各自不同的特征。对道路网络中每个部分的改进情况进行记录是一项非常有意义的工作，这些改进内容既包括道路的通行能力，也包括道路不同区域的使用者方便性等。〉见图 37

118

图 36：
土地的混合使用结构——调查区域土地使用相
协调还是引起了土地使用冲突？

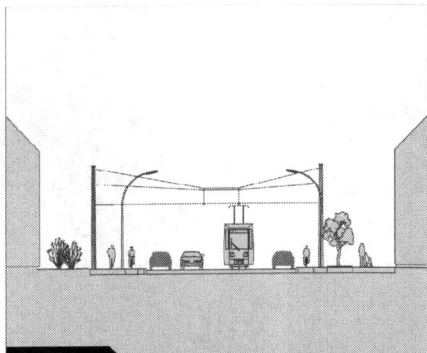

图 37：
可采用标准断面图对道路的空间特征进行表达

— 道路与相邻建筑的宽度比值；

— 将道路空间进行车道和路侧区域的划分；

— 道路空间的类型、形式以及尺寸；

— 道路装饰和表面材料；

— 照明设备与类型。

交通结构的进一步深入分析需要专业工程师的参与。在进行城市
分析之前，可能已经有现成的交通分析报告供参考。除交通类型（客
运或者货运）之外，交通需求（货运、专业运输或者休闲旅游等）、
交通距离（源交通、目标交通、内部交通以及中转交通等）等相关
因素均可能对交通压力造成影响。

交通设施可以根据其不同的设施结构特征进行多种不同的分类。
而在城市分析中，一般主要是根据交通设施使用者的类型进行分类，
比如机动车交通设施、区域公共汽车交通设施以及人行交通设施或者
自行车交通设施等。

提示：
    交通压力可以通过交通车辆出现的间隔时间
计算得到，一般最常用的是每日的平均交通量。
该数据可以针对单个道路横截面进行记录和计算
得到。所得的结果图形能够反映某一道路截面在
一天中不同时段或者一周中不同天的不同交通
压力。

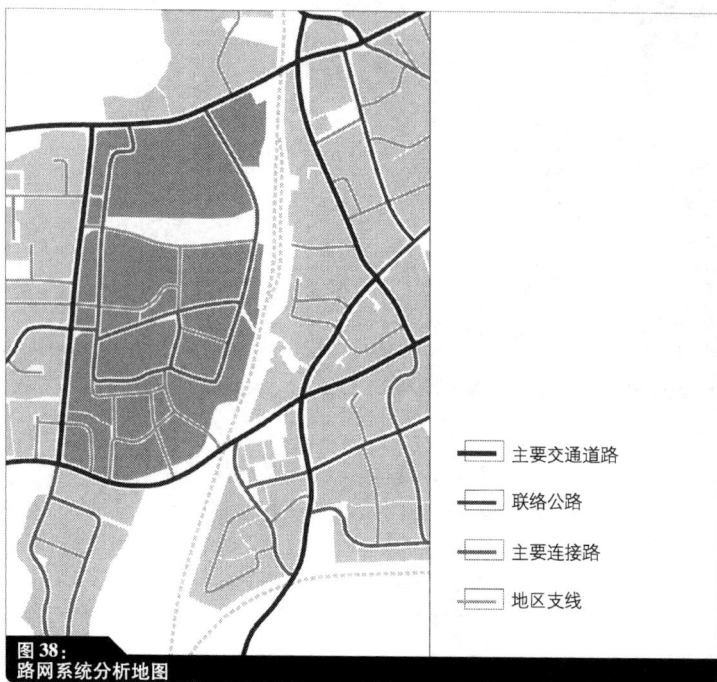

图例：
主要交通道路
联络公路
主要连接路
地区支线

图 38：
路网系统分析地图

P49
动态交通

机动车交通设施

进行动态交通分析的第一步是对所在区域的道路根据其通行能力进行分类。〉见图38

— 高速公路/干线道路；

— 主要交通道路；

— 联络公路；

— 主要连接路；

— 地区支线；

— 其他重要交通道路；

— 农田或者林间道路。

这种分析方法将确定一个区域的交通路网体系结构。大多数情况下，所有的道路能够形成一个层次型网络。比如，路网中的交通车辆是不均匀分布的。道路类型的划分基于其分工的不同，交通车辆在使用率较高的公路中得以集中，在使用率中等或者较低的公路中得以疏

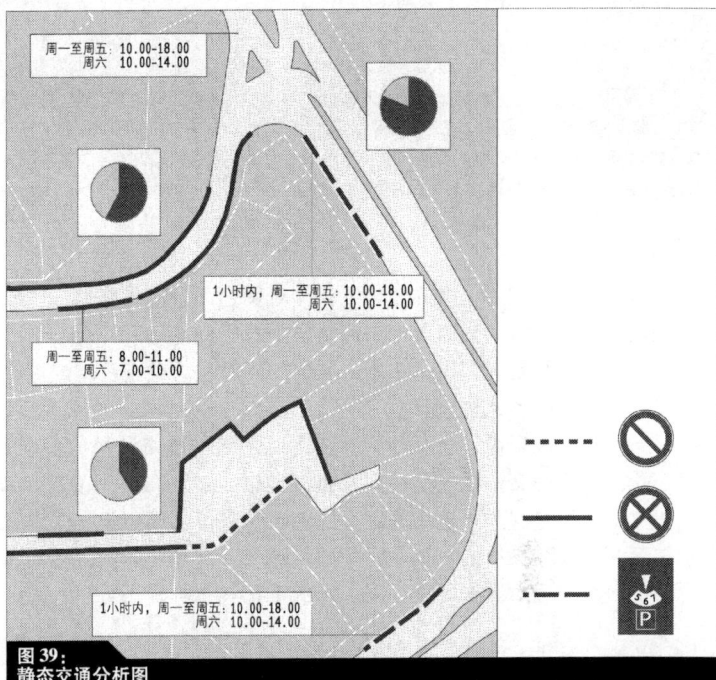

周一至周五：10.00-18.00
周六　10.00-14.00

1小时内，周一至周五：10.00-18.00
周六　10.00-14.00

周一至周五：8.00-11.00
周六　7.00-10.00

1小时内，周一至周五：10.00-18.00
周六　10.00-14.00

图39：
静态交通分析图

散。而交通路网体系中的组成部分可以根据其交通压力的不同将功能划分为：联络路（交通压力大）、支线（交通压力中等）、停放道路（交通压力小）。当实际情况与计划发生背离的时候，可能引发相应的交通冲突。

静态交通

　　停止的车辆（以及无功能的车辆）被归为静态交通。在城市分析中一般会对公共停车场所的比例和使用强度进行标注，一般可以在公共路网以及特定设施（比如地面停车场、地下停车场或者多层停车楼等）中设置停车场所。这些停车场所的管理方式也存在不同，包括免费停车场、限时停车场以及付费停车场等。〉见图39

　　调查区域的不同土地使用性质对停车场所的空间需求不同。也就是说，确定停车场所的位置和所占空间比例时，需要考虑土地使用性质的因素。

提示：

交通路网也可以根据其投资者的不同来进行分类，比如高速公路、国道、省道、地区道路、局部道路以及私有道路等。但是，采用这种方法不足以对一个较小调查区域的道路进行区分。

提示：

停车场所的使用强度的分析与对应的不同时间段相关，包括分析占用的停车位与总停车位之间的比值。通过占用率图表能够反映停车场在一天或者一周的过程中使用程度的变化情况。

调查区域之外的背景研究

交通结构的分析不仅要对调查区域内的路网进行研究，还需要就该区域的交通设施如何与整个城市或者地区的交通系统形成一个体系进行研究。这意味着，我们需要对包含调查区域的更广范围的交通路网的形式进行分析——比如，放射状路网体系或者环形路网体系等。一般来说，会有若干不同交通路网与调查区域的不同地区发生重叠。远程交通系统是通向调查区域以外地区的一种直接的交通结构形式，但是这种系统为其经过的地区带来了交通负担，同时也增加了尾气排放和噪声的发生。在交通路网或者交通功能发生空间重叠的时候，可能会对调查区域带来一定的干扰。〉见图 40

P52

区域公共交通设施

区域公共交通设施指的是针对所有人开放的区域短程交通设施，这种交通通常有固定的路线和时刻表。对区域公共交通设施分析主要是针对车站的实用性、服务性以及路网的结构进行调查。区域公共交通系统可以做以下分类：

— 有轨系统（区间列车、远郊列车、城市轨道交通、地铁、有轨电车等）；
— 定期公共汽车系统；
— 其他交通系统（比如，出租车、市区巴士等）。

站点的设置是一个非常重要的因素。较好的路网系统中站点距离居住地或者工作场所的距离不超过 300m（直线距离），而 500m 之内则是一个可以接受的水平。在进行该距离分析的时候，需要将地形、已有人行通道设施（比如，行人交叉路）以及需要绕行的空间障碍物等因素考虑在内。同样，相应的改进设施（比如，为平台所设置的电梯或者与高路缘相对应的客车低地板技术等）也能够对通行能力造成影响。

区域公共交通体系的质量很大程度上取决于站点的服务设施以及路网的运行速度。在进行区域公共交通分析时，提高车次的准时性、

图 40：
整体交通分析图

主要交通道路

联络公路

铁路

换乘的方便性，增加运营时间和提高公共交通的运行速度（比如，划分公交专用道、通过交通灯指示赋予区域公共交通通行优先性等）是提高区域公共交通质量的重要措施。〉见图41

调查区域之外的背景研究

调查区域可能包含在一个集中路网或者分散化的路网之中。在集中路网中，乘客只能通过某个特定的中转点转向另外的路线或者路网中；而在分散化路网中，则存在多个中转点。然而，多数路网则或多或少介于两种形式之间。所以，在进行区域公共交通体系评价的时候，这些中转点的空间邻近性与其所具有的站点的数量同等重要。

P54

人行交通及自行车交通设施

根据调查区域交通设施水平的不同，人行交通和自行车交通设施可能会进行单独设置，或者仅作为交通路网的一个部分进行设置。

只有在针对独立的人行道或者进行非常详细的交通结构分析时，

图 41:
短程公共交通分析图

才会对独立的人行道或者道路两侧的人行道交通进行分析。

通常情况下，自行车交通分析需要对所有自行车交通设施进行记录，并对道路两侧的自行车道、自行车运动区（从普通车道中划分出的有标志路线）、独立自行车道以及自行车路加以区分。

人行交通和自行车交通路网以及安全系统的优化是确定其对调查区域不同地区以及周围地区联系质量的主要因素。因此，只要是与自行车交通相关的所有节点设施均应该加以记录，而不是仅仅记录安全路口（比如道路交叉口以及其他路面交叉口），并对路网系统中特别不便利或者危险的区域进行命名。另外，调查区域中不论土地使用性质如何，只要有任何类型交通出入口存在，均应该加以记录。〉见图 42

调查区域
之外的背
景研究

由于人行交通以及自行车交通将调查区域和周边环境或者邻近区域进行了相互联系，在城市分析中对这种联系的研究具有重要的意义。在进行整个城市或者具有一定规模的区域分析时，一种较好的方

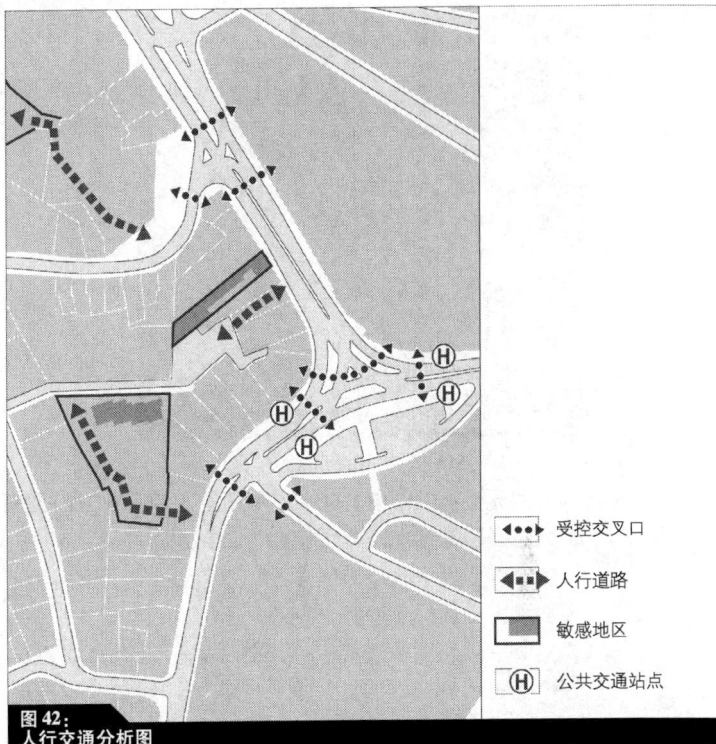

图42:
人行交通分析图

图例:
◀••▶ 受控交叉口
◀■■▶ 人行道路
▨ 敏感地区
Ⓗ 公共交通站点

法是将人行交通和自行车交通系统划分为休闲路网和日常路网进行分析。休闲路网一般通向一些地区休闲设施，而日常路网则要求能够尽快到达城市活动的目的地。

交通结构分析中观察结果总结

土地的不同使用性质具有不同的出入需求——比如，对交通工具以及交通设施的尺寸要求不同。因此，需要根据投资和出入口设施质量要求来确定相应设施的档次和对周边的干扰程度。同时，调查区域中土地的不同使用性质具有相应不同的防尘和防噪声需求。具有较大交通压力的机动车交通系统将对邻近房屋的环境质量以及市区通行的时间消耗造成影响。另外，机动车交通系统还相当于空间屏障，而将调查区域进行了划分。同时，道路中过大的交通压力能够标示城市生活的中心所在。在进行总结评价的时候应该体现出以上所有的主要因素。

图 43：
道路评价需要考虑许多不同的相关因素

交通类型的不同可能会对城市空间的需求方面带来冲突。某一个交通路网的高标准设施可能会为该路网另一个设施的功能带来影响。打个比方，大量的公交车道可能会为公共交通使用者带来便利，但会影响私人交通的通行能力。

所有的路网体系都应该评估其是否具有容易辨认和记忆的空间方位指示设施。好的方位指示设施能够增加使用者的舒适度和安全感。〉见图 43

P56

城市空地和绿色结构分析

城市空地和绿色结构分析主要包括以下两个主要任务：

1. 表示出城市空地的自然环境基础设施，并根据物种保护和生态法确定它们的受保护需求。城市分析人员通常不需要自己进行生态分析，而是从其他专业分析报告中获取分析区域的相关信息。

2. 对城市空地的使用、质量感官质量以及城市空地与房屋结构的相互关系进行分析。城市空地和绿色结构的空间质量需要城市分析人员亲自进行调查和评价。

空间相关特征

与土地使用结构分析相似，城市空地和绿色结构分析需要对城市空地的所有真实使用情况进行记录。与土地使用结构分析不同的是，该分析只需要对调查区域内部进行分析。由于分析的主题会与土地使用结构分析发生重叠，所以在进行城市分析的时候一般会省略这个耗时的过程，而只是对城市空地和绿色结构分析所得的主要

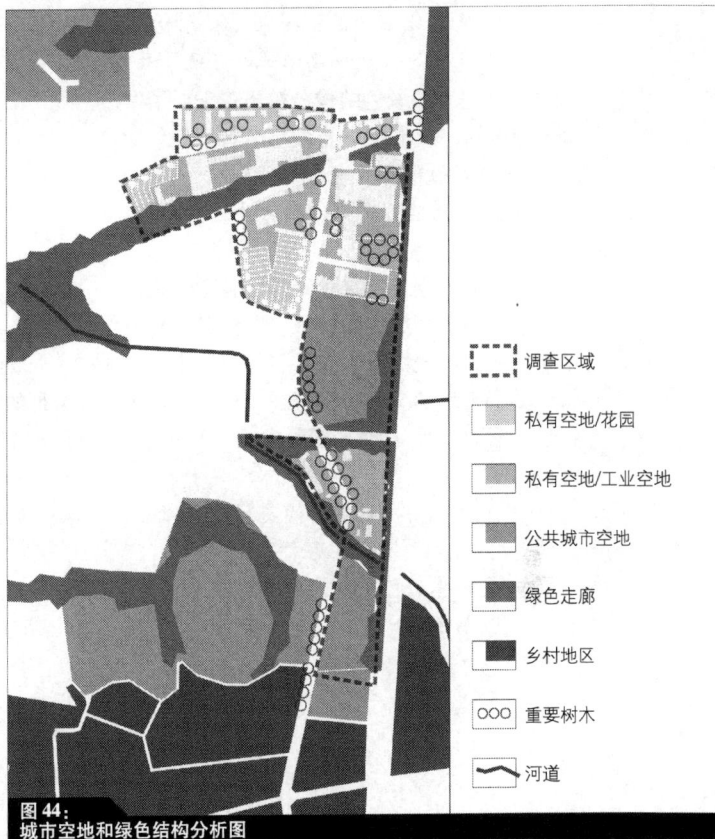

图 44:
城市空地和绿色结构分析图

图例:
- 调查区域
- 私有空地/花园
- 私有空地/工业空地
- 公共城市空地
- 绿色走廊
- 乡村地区
- 重要树木
- 河道

特征进行简单分类。这些空间一般首先根据其所属关系进行分类和记录。〉见图44

- 公共城市空地，比如公园、集市场所、河边散步场所、水体等；
- 相关设施所配套的公共城市空地，比如学校、幼儿园操场、动物园等；
- 半公共城市空地，比如公寓群内部院落、办公楼空地等；
- 私有城市空地，比如花园、公司场所等。

针对整座城市进行城市分析时，对地面景观分类通常仅考虑城市空地的类型，一般采用如下分类：

- 耕地（比如农田、牧场、葡萄园等）；

— 林地（比如受保护林地、休闲林地等）；

— 其他空地（比如高沼地、湿地、水体等）。

空间表现的
结构元素
城市空地和绿色结构分析不仅记录了城市空地，而且还记录了对周围空间有一定影响的自然结构特征，包括：

— 显著的独木和树木群；

— 成排结构，比如道路两旁的树木或树篱等；

— 绿色走廊。

由于城市空地和绿色结构的独立性很强，一般很难通过某种标准方式表示出他们的空间特征。在记录分析结果的时候，一种较好的方法是采用空间概图或者摘要符号进行表示。见图45

特征评价
城市空地和绿色结构分析的第一步是对个体特征的内在本质进行评价。需要对具有如下五种中的一种或者多种重要特征的元素进行评价：

— 历史重要性：重要的花园建筑或者历史文化元素（比如历史公园、文化景观元素等）；

— 空间体验性：空间性质，比如种植用地与城市空地之间的相互影响、城市空地与城市地貌之间的累积作用等；

— 休闲性：具有特殊休闲功能的地区（比如，公园以及居住区附近的树林等）；

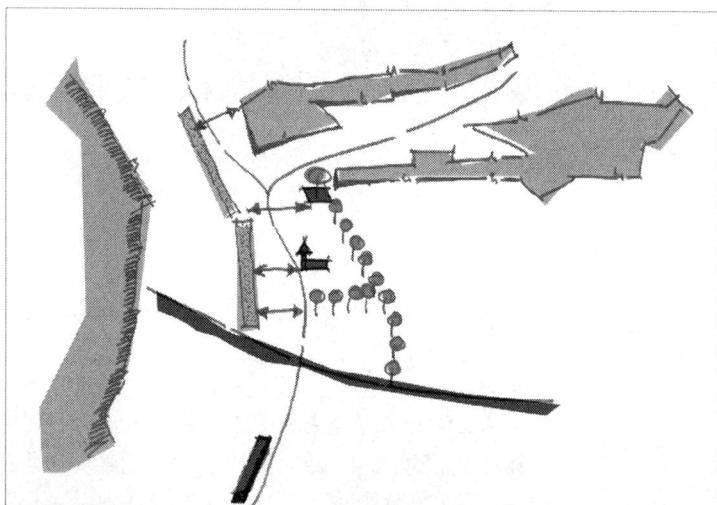

图45：
采用概略手绘图进行城市空地的综合表达

提示：

城市空地的休闲性根据土地使用强度的不同而异。通常，居住区附近的城市空地会有大量的休闲人员，而从比例上来看，乡村地区的人数则很少。

— 功能性：城市空地和绿色结构的使用与城市地表使用的一个重要部分；

— 保护性：法律规定对相关城市空地和绿色结构元素的保护。

保护区与保护元素

根据相关规划文档和专家报告中得到的保护区域保护元素清单必须整合到城市分析中：

— 禁止某些使用功能的保护区：动植物保护、地面景观保护（比如保留地区和受保护景观等），以及饮用水资源保护、空气净化走廊等；

— 受保护的基础元素：为了动植物生存所进行的个体元素保护，比如独木、树篱、河岸等。

在进行城市分析时，与调查区域相关的所有保护区和受保护元素均应该准确地标注在地图中。〉见图 46

调查区域内部的相互影响

在进行城市空地分析时，最有意思的一部分工作是进行城市空地与房屋结构的相互影响分析。通过城市空地汇总表，能够得到不同类型城市空地的空间分布以及所占地表面积的比例。通过这些数据计算，能够帮助我们确定调查区域特定休闲区域的设置，通常设置的依据是调查区域的人口数量。

城郊的小规模房屋通常都具有各自的花园，所以对城市空地空间的需求较低。而城市中心的密集大规模房屋则对城市公共休闲空间存在较大的需求。

调查区域之外的背景研究

在进行城市空地和绿色结构分析时，需要将调查区域整合到周围环境和整个城市中，作为整体进行分析。其中需要重点考虑的是独立绿色空间与绿色基础设施之间的空间联系，以及调查区域城市空地与周围环境的相互联系。

城市空地和绿色结构可能会按照一定的几何形状分布，比如采用圆环状或者中心走廊形式。而不同的几何形状通常按照一定的形式（比如，同心形式），发生重叠。〉见图 47

图 46：
保护区分析图

　　此时，较小的连接元素和较窄的绿色走廊也可能比较重要。在进行整体分析的时候，同样需要分析是否有调查区域之外的城市空地能够为区域内的居民提供休闲场所。在分析时，需要记录以下信息：

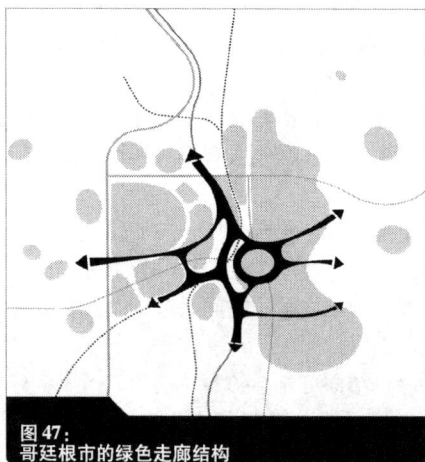

图 47：
哥廷根市的绿色走廊结构

—　房屋与地面景观的整合、房屋与城市空地的连接、房屋附近休闲场所的入口；

　　—　绿色连接元素，比如林荫道、绿化边缘等；

　　—　城市空地与城市地貌的相互作用，比如绿色河岸、植被堤坝等；

　　—　景观轮廓，比如开敞式或者封闭式景观。

　　许多城市空地和绿色结构的性质只有在相互作用时才得以显现。所以城市空地和绿色结构的表达通常是不完整的，中间存在缺口以及空间障碍。从整体规划的角度出发，在进行城市分析时存在的一些联络线和长廊需要被记录和表现。

P62

建筑物和房屋结构分析

　　建筑物和房屋结构分析任务包括：对调查区域的空间组成进行分析，对开发特征进行描述，以及揭示建筑物结构与城市空地之间的相互关系。

个体特征记录

　　不同的建筑物和房屋结构分析具有的详细程度差异非常大。在进行逐个小区分析时，需要按照标准分类，将每个单独建筑的结构特征记录到地图中。对于重点考察历史重要性特征的城市分析，必须达到一定的详细程度，可能需要记录如下的结构特征：〉见图48

　　—　建筑的楼层和尺寸（建筑密度）；

　　—　建筑类型（比如联排住宅、街区建筑、带状建筑等）；

　　—　建筑形式（比如封闭式或者开敞式建筑形式）；

　　—　屋顶形状（比如坡屋顶、斜脊屋顶、平屋顶等）；

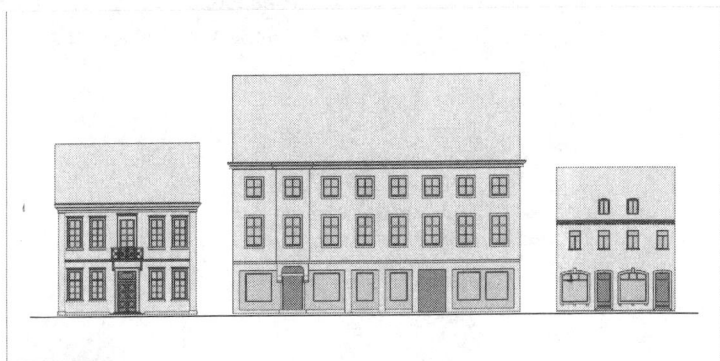

图48：
在进行单独房屋的评估时，通常将建筑的主立面作为重要特征进行记录

— 建筑方向（比如房屋正面面朝街道或者山墙面朝街道等）；

— 入口系统（比如具有楼梯口的公寓、具有外部走廊入口的街区或者具有内部入口的街区等）；

— 附加风格元素（比如装饰物、装饰梁等）；

— 建筑式样（比如建筑各面一致或者设有正立面等）；

— 房龄，建造年代；

— 所有权关系（租用或者所有权类型，多人所有或者多人经营等）。

颗粒化分析　　在进一步细分、进行相似结构特征分析，或者将调查区域按照房屋结构形式和最初建造阶段进行划分时，需要采用逐个小区分析的方式。此分析通常根据建筑规模、建筑形式以及建筑物房龄来进行。对于出自同一城市规划的调查区域（比如，处于同一发展区域的细分）最容易进行颗粒化分析。常见的建筑发展结构包括：

— 封闭街区发展建筑；

— 开放式带状发展建筑；

— 密集地面建筑；

— 独立公寓综合体；

— 家庭住所发展建筑。

一般很难对结构形式明显混乱、不够统一的地区作以上分类，而调查区域一般由这些地区组成。在这种情况下，通常采用地区的一致性程度作为分类依据。比如，可能将不同地区做如下划分：

— 20世纪末、21世纪初以来具有密集型街区结构的建筑；

— 20世纪末、21世纪初、20世纪50年代或者60年代以来具有大型密集型街区结构的建筑；

— 不分年代、所有具有开敞式发展结构的建筑。

提示：

　　关于城市发展类型的更多信息可以参考本套丛书中的托尔斯腾·别克林（Thorsten Bürklin）和迈克尔·彼得莱克（Michael Peterek）所著、张路峰翻译的《城市街区》一书（中国建筑工业出版社，北京，2011年）（征订号：20433）。

图49：
根据建筑物年代和特定房屋结构进行的区域划分

由于对建筑物的单独评估工作需要消耗大量的时间，在建筑物和房屋结构分析中通常省略这个过程。对于这种情况，通常是在进行分析时首先将具有相似结构的建筑物进行分类。对于具有混合结构建筑物的城市，要求进行这项工作的分析者具有充分的实践经验。〉见图49

空间体验　　城市空间由大量建筑物或者大量带有城市空地的建筑物相互作用而形成。城市空间具有不同的性质，比如：〉见图50和图51

　　—— <u>界线</u>：封闭式空间或者开敞式空间的界线（明显的分界线或者逐渐过渡）；

　　—— <u>几何形状</u>：由具有一定几何形状的边界形成的空间（比如，建筑轮廓线）或者由不规则布置的建筑形成的空间；

　　—— <u>比例关系</u>：反映建筑物与城市空地的范围、密度等关系。

　　另外，也需要对城市空间中存在的缺陷以及干扰程度进行分析。

133

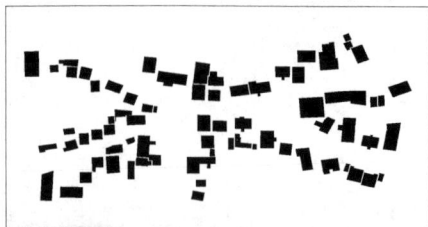

图 50:
具有不规则布置的建筑物的开放发展结构所形成的城市空间

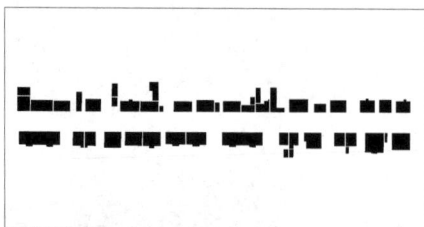

图 51:
占主导的封闭发展结构具有一致的建筑轮廓线

调查区域
之外的背
景研究

　　在进行建筑物和房屋结构分析时，需要考虑一些延伸至调查区域之外的空间结构或者结构元素的影响。而这种分析通常能够揭示房屋结构的本质关系，同时有助于城市内的导航。一般的关系包括：

　　　— 地形与房屋结构的关系：根据不同地区的地形差异采用不同的房屋结构（比如，平地上采用大型建筑，非平地上采用较小规模的建筑）；

　　　— 中心区域的定位：在城市中心具有某种特征的建筑得以增加（比如，城市中心建筑物密度相对城市外围较大）。

　　以上关系相违背的做法通常被认为是一种干扰，但有些特别显眼的建筑或者空间元素也可能是重要的地标建筑。

总结

　　建筑物和房屋结构分析提供了关于调查区域使用功能相关的信息，所以在进行该分析的时候，最好能够与土地使用性质分析进行对比。通常来说，具有混合土地使用性质的市内密集城市邻里承担着为市内周围地区提供城市功能的任务。然而，由于这些区域具有密集的平面规划，相应对公共休闲区域的需求也更大，所以这些休闲区域可能需要由调查区域之外的地区来提供。

P65

社会空间分析

　　社会空间分析关注的是具有不同人口结构的城市空间的相互作用关系，并根据人口和空间结构特征的相对一致性对城市区域进行划分。社会空间分析工作一般由社会学家来进行，并和城市分析家一起将分析报告中的重要结论与其他城市分析的结构进行对比。这个过程能够为城市社会空间规划提供有价值的信息，并从空间和时间上找出影响城市发展和普遍的社会利益的关键点。〉见图 52

工作方法　　　　没有针对社会空间分析的特定方法，可能存在两种不同的提问方式，也就是相应的两种方法，如下：

　　　　——人口特征的分布情况如何？

　　　　——某一特定区域存在怎样的人口特征？

对已有评估
结果的参考　　　　实际上，城市行政管理者将对城市内的人口结构和发展情况进行持续的调查。通常情况下，各个人口统计分析区域是根据行政区划范围来确定的。然而，对于城市分析而言，调查区域一般与人口统计分析区域并不相同。所以对调查区域而言，参考已有的调查结果不能得到准确的结果〉见图53。为了对调查区域进行分析，需要从相应的人口普查部门获取相关数据。也就是说，在开始分析的时候需要对数据

图52：
人口统计学分析：基于小区进行的人口统计分析

图53：
城市分析的调查区域与人口统计分析区域通常
是不相同的

的可用性、技术的可实施性以及数据的保护进行相关讨论。

　　进行社会分析的最简单的形式是进行人口情况登记。通过对调查对象的相关信息进行记录，可以分析得到人口发展情况。

　　通过计算相关空间单元的居民数量，可以确定相应的密度数据——人口密度（每公顷土地所具有的居民数量）或者居住率（每户家庭中的平均人口数量）。

　　根据调查方式以及人口调查部门的能力不同，可以对调查区域的若干特征进行分析。相关的人口特征可能包括年龄、性别、婚姻状态、子女数量、出生情况、死亡情况、国籍、出入境情况以及社会福利需求等；而空间特征可能包括家庭数量、规模或者建筑年代等。

　　为了研究发展趋势，需要进行长期分析，所以应该每隔一段时间就对上述的特征进行记录分析。

　　通过各个街道或者建筑物等微观层面的分析，能够反映出不同人群的不均匀分布情况。如果调查区域的一些部分具有显著不同的城市规划结构，则需要对这些地区的社会特征和空间特征之间的相互作用进行更加细致的分析。在分析中可能会用到以下两种不同方法：

　　1. 将调查区域的所有特征作为一个整体，与其他调查区域的相关特征进行比较；

　　2. 根据调查特征的不同，将调查区域进行空间划分——比如，确定调查区域的哪些部分具有某种特征。

　　在将调查区域的情况与其他区域（比如，行政区、整个城市、某个地区等）进行比较的时候，一般意味着需要从更高级的机构中获取相应的数据。在使用来源不同的数据时存在一定的风险——由于数据采集方法的不同，可能导致相互比较的结果无效。

总结　　根据对调查区域的分析结果，一般能够得到与生活在调查区域内人口的社会状况、调查区域与周边社会环境的相互联系等相关的结论。由于通过单个特征所提供的信息很难得到上述结论，所以对调查结果的简单理解可能会导致错误的解释。比如，高的人口波动性可能预示着人口稳定性较低，所以说居民对邻里的熟悉程度较低。但是，对于主要被学生占据的地区，这种结论不攻自破，而且这种情况很常

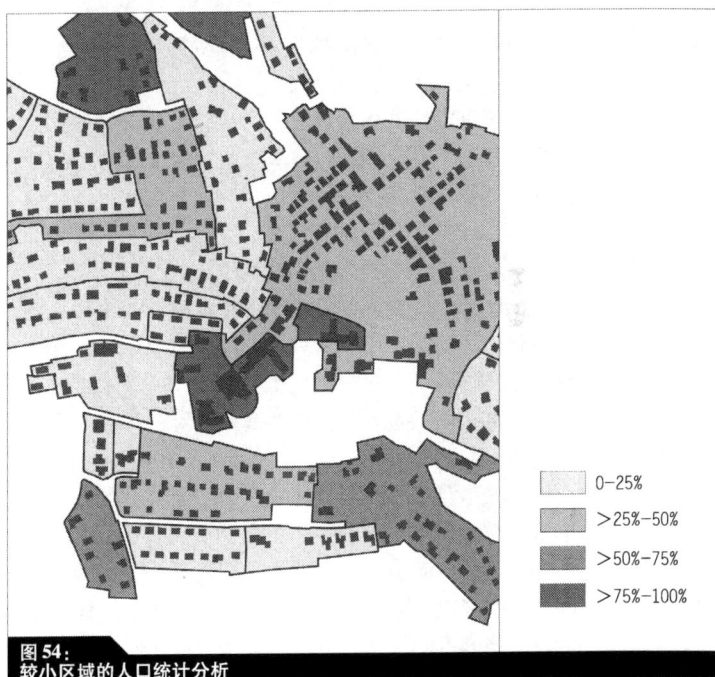

	0–25%
	>25%–50%
	>50%–75%
	>75%–100%

图54：
较小区域的人口统计分析

见。所以，在做出任何结论之前，应该充分考虑若干相关的特征因素以及调查区域的预期功能。一种特别好的做法是，将社会空间分析的结果与建筑物和房屋结构、土地使用性质等分析结果进行比较，同时还应该运用居民和人口活动相关的专业知识（参考"在分析对象地区进行的工作"一章相关内容）。

社会种族隔离

社会种族隔离从空间上对人口结构进行了疏远，与社会混合种群指导原则背道而驰。然而，由于社会分化的原因，意味着不同的社会环境和社会阶层（比如生活方式相似的人群）之间一直存在着一种相邻关系。由于特定的社会环境能够使得人们更加容易地融入特定的社会网络中，所以这种人口发展结构并不一定存在问题。但是，空间上的极端集中则会导致社会分离区的形成，预示着在同一社会邻里中产生了极端相对的集中情况，以及社会公益服务和现代化方面的分离。在进行社会空间分析时，必须明确这些区域的存在。

分析结果的解释与表达

最初的评估结果由各个分项分析得到，而并不考虑各部分分析之间的相互关系。

由于可以根据标准参数进行比较，所以对各个分项分析结果的评估相对比较容易。通常，一项分析需要根据以下两项内容进行评价——结构和功能。

分项分析结构之间的相互影响

比如，一座城市能够看作各种空间、功能和社会影响相互作用形成的整体。同时，由于对城市空间的需求矛盾将导致许多城市功能障碍和冲突的发生。在进行城市分析时，一项重要的工作是区分哪些冲突属于可接受的范畴，而哪些属于实际的冲突。不论是对于整个城市还是城市中的一部分，均不可能对所有的相互作用加以记录。所以说，在城市分析中对各部分之间相互影响的分析与各分项分析一样，都是对整个城市进行的一个抽象性模型表达。

分项分析

分项分析的结果评估通常采用的是一种因果关系方式。比如：空间环境的产生原因是什么？但是，更多评估工作则是建立在各分项分析中不同因素的相互关系之上。比如：交通结构分析结果表明，需要增加某条道路的通行能力。一旦需要考虑分项分析之外的其他因素——比如，该道路位于居住区——那么可能就无法满足相应的需求。

通过分项分析的比较，能够确定调查区域的特定需求以及某种城市功能的不足或者过剩。这种比较通常需要涵盖调查区域之外的地区。通常，主要需求因素如下：

— 休闲场所的发展密度和设置；

— 人口以及居住配套设施的设置，比如提供日常必需品的配套设施等。

总结评估

在进行总结评估时，必须明确最初进行城市分析的出发点（参考"准备阶段和设计阶段"一章相关内容）。总结评估必须对最初的问题进行回答，或者对调查区域的情况给出一个全面的描述。对最初问题的回答通常并不是指回答调查区域的功能情况，而是回答调查区域在整个城市规划中所扮演的角色——确定城市某个新的发展计划是否合适。分析家则在此基础之上决定土地合理的使用性质。另一方面，如果分析针对的是整体情况而不是某个特定主题，则需要明确相应的优点和机会以及缺点和危险。对于这种情况，在对调查区域的相关因素进行评估之后，需要提出新的功能建议。此处的"功能"指的是调查区域在背景城市的更广范围内所扮演的主要角色。比如，调查区域可能单纯作为住宅区，也可能通过配备相应的设施，为整个城市提供某项功能。调查区域的功能必须在将整个城市作为背景的前提下确定——比如，调查区域的特征必须通过与整个城市的结构特征对应分析后才能得到。

在对城市功能障碍进行总结分析时，需要对调查区域的功能不足和本质不足进行区分。功能不足指的是影响调查区域发挥应有功能，而本质不足则指的是调查区域的构成部分对人们的健康居住、工作条件有害，或者是对人们的生活、工作造成了危险。对于调查区域，可能只有少部分存在本质不足。

对于机会的危险的评估需要基于现状之外，通过调查区域的分析结果，确定其将来可能的发展情况。分析家必须根据相关理论，对调查区域特别是空间发展趋势进行分析。打个比方，机会可能是将城市荒地改造为绿色空间，或者将闲置建筑改造成新型房屋等。而危险可能包括不利趋势的加剧或者遭受城市规划失误等。

实际上用来表达以上因素的方式有两种：

1. 采用评估矩阵，比如将利弊因素放在统一表格中进行分析。而表格中的不同因素可以根据其重要性进行排序。〉见图55

优点	不足
– 出入便利； – 绿色空间充足； – 文化多样性。	– 局部公共设施缺乏； – 噪声大； – 土地闲置。
机会	危险
– 适合年轻家庭居住； – 闲置房屋的类型转变。	– 闲置情况增多； – 局部形象的破坏。

图 55：
评估矩阵中的利弊分析

2. 采用图形方式进行利弊因素的表现。这种方法在进行空间相
   关因素表现时特别有效。〉见图 56

P74

实施方案的确定

如果通过城市分析给出了调查区域的整体情况，那么接下来将对
所有的分析结果进行考虑，并为城市将来的发展情况建立一个参考模
型。由于存在许多设计者所无法控制的因素，所以实际的发展并不是
完全遵照规划展开的，建议采用以下两种方式：

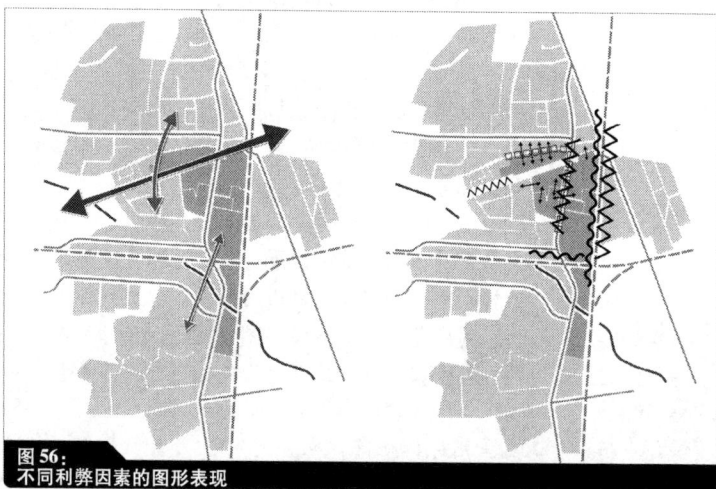

图 56：
不同利弊因素的图形表现

140

1. 建立调查区域的发展模型；
2. 给出可替代的草略方案或者设想发展模式。

参考模型　　参考模型作为一个整体战略目标，可以为未来的每一个发展步骤提供参考方向。所有可行的单个或者系列措施的实施都不能忽略参考模型，而该参考模型是在城市分析结果的基础之上，按照以下三条原则所建立的：〉见图57

— 描述性或分析性：根据研究所的趋势确定未来发展情况；

— 目标化或标准化：根据设想目标确定所需措施；

— 原始性或调查性：研究可能推动未来发展的相关因素。

根据城市分析所设想的目标，可以通过以下三种方式对调查区域的发展情况造成影响：

1. 活跃或者推动作用

— 工程准备，比如可行性研究；

— 实施计划或者应变计划的制订；

— 确定可能地区；

— 寻找财政支持。

2. 控制作用

— 为工程实施制定相关规则；

— "假定"选项制定。

3. 保守或者被动作用

— 避免不期望的发展情况；

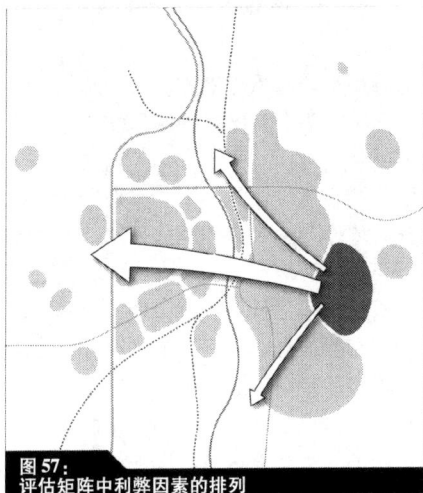

图57:
评估矩阵中利弊因素的排列

— 保护已有的可能地区。

分析结果的表现

　　为了明确表现空间应用的不同方面，在可能的情况下，应该采用图形和文字的双重方式来表现评估结果及其解释。将所有的分析结果输入到比例图中，同时给出相关数值和图示。根据详细程度的不同，将记录特征分为标题和次标题进行记录可能是一个较好的做法。图示应该能够反映不同的区分情况，将统一标题下的分类用同一种颜色进行表示——比如，采用红色灰度梯度。规划图和地图中的表示方法应该基于相同的图示。根据所需的表示类型不同，可能会用到建筑物、建筑小区或者建筑场地的轮廓线。〉见图58

　　虽然城市分析中没有统一的标准规划符号，但正式的规划文字中对所用到的规划符号有明确的说明。对分析结果进行简单易懂的图形表现是城市分析中的一项重要工作。

　　在将分析结果表现到地图上的过程中，应该根据调查区域尺寸和分析详细程度的不同确定合适的地图比例。其中的决定性因素是需要表现的城市空间元素的最小决定单元（比如，空间单元）。

　　对于整个城市的分析，其最小的空间单元通常是建筑场地或者小区，所采用的比例尺在1:20000到1:5000之间。一般采用地形图作为背景。

　　地区层面的调查通常包含单个建筑物或者小区的信息，所采用的比例尺在1:2000到1:1000之间。一般采用房屋图作为背景。

　　小型地区调查（比如单个建筑小区或者街道）通常采用的比例尺在1:500到1:200之间。

　　略图和图标适合用来表示重点地方。通过简化的含有其重要的结构特征的略图或者图标，为分析结果的阐释以及空间的表现提供了一种灵活的方法。〉见图59

　　统计分析可以将调查区域作为一个整体来进行，并与含有其他人口普查区域的调查区域作为一个整体来进行比较，或者对调查区域的不同地区进行分析。

　　统计数据的分析通常采用计算机电子表格程序进行，这些程序能够将统计数据进行图形化表现。如果只是将调查区域作为一个统计对象，那么采用这些程序就足以完成评估结果的表现。如果将对调查区域进行详细的调查并划分成了几个不同的人口普查区域，那么还需要采用地图来表示相应的人口普查区域。而评估结果通常是通过数据表格的方式进行整理，并采用标准图示的方式表现在地图中。〉见图60

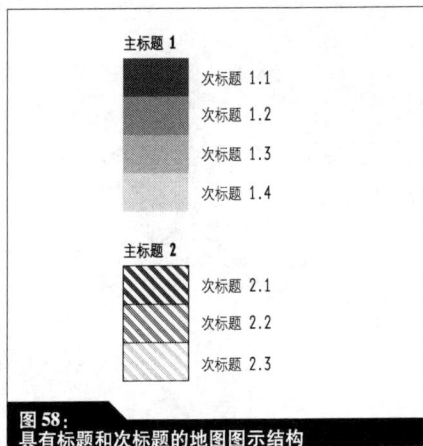

图 58：
具有标题和次标题的地图图示结构

图例：
主标题 1
次标题 1.1
次标题 1.2
次标题 1.3
次标题 1.4

主标题 2
次标题 2.1
次标题 2.2
次标题 2.3

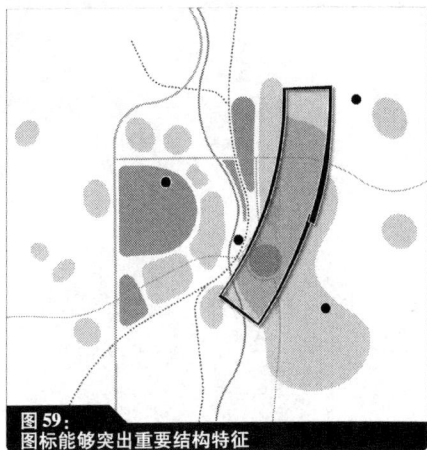

图 59：
图标能够突出重要结构特征

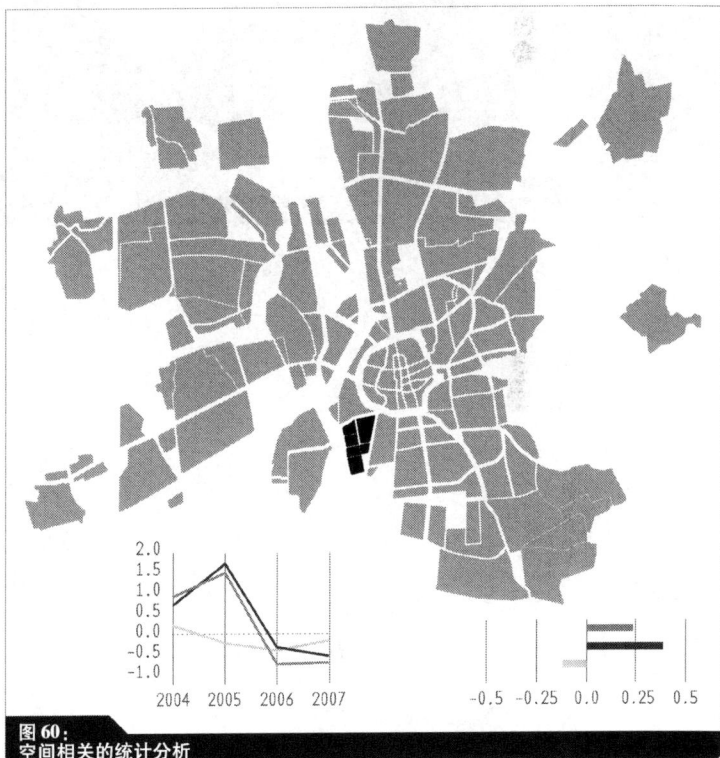

图 60：
空间相关的统计分析

图 61：
在进行城市分析的时候需要考虑对调查区域的感知上的不同

结语

城市分析并不是将若干分项分析的简单相加,而是将分项分析整合成一个整体。城市分析的质量取决于对每个调查区域中不同相互影响因素的意识,而不是将不同分项因素的罗列。只有在充分考虑大量不同因素的相互影响之后,才能对分析区域有一个清晰的整体性认识,而且影响城市分析最终准确性的原因也在于此。

对于单个因素的信息过度收集将对相互影响的分析带来阻碍。所以,在进行分项分析的时候,应该从分析将有效处理结果的角度出发,确定合理的详细程度。空间信息和人口相关信息则存在大量相对可以很快获取的数据,为了避免获得大量无关的数据,在数据收集之前必须进行相关性检查,而这项工作需要分析者的经验以及与分析目标相关的意识。

城市分析为分析区域的未来城市规划措施提供了一个抽象性的模型表达。然而,由于城市分析并不能满足某些方面的研究,比如感情联系、社会习惯等,所以城市分析中常常会遇到不同群体在城市感知上的矛盾——所以进行城市分析的时候,应该考虑到这种不可避免的因素。

城市分析中对于居民的愿望咨询工作,是居民、政府以及城市生活中的其他参与者能够和谐相处的基础。在城市发展越来越多地建立在已有城市结构基础之上的背景下,采用这种城市分析是城市可持续发展的一个重要部分。

附录

参考文献

Christopher Alexander, Sara Ishikawa, Murray Silverstein: *A Pattern Language: Towns, Buildings, Construction*, Oxford University Press, 1978

Carl Fingerhuth: *Learning from China. The Tao of the City*, Birkhäuser Verlag, Basel 2004

Peter Hall: *Urban and Regional Planning*, Taylor & Francis, 2002

Kevin Lynch: *The Image of the City*, MIT Press, 1960

Franz Oswald, Peter Baccini: *Netzstadt. Designing the Urban*, Birkhäuser Verlag, Basel 2003

Colin Rowe: *Collage City*, MIT Press, 1984

Aldo Rossi: *The Architecture of the City*, MIT Press, 1984

Thomas Sieverts: *Cities Without Cities. Between Place and World, Space and Time, Town and Country*, Routledge Chapman & Hall, London New York 2003

图片来源

图7：© kittel+partner, Dresden. Shows: Peek&Cloppenburg department store in the centre of Chemnitz

图9（左）：Graphic: G. Schwalbach. Source: The city of Göttingen in the Middle Ages, Map 2 from O. Fahlbusch: "Die Topografie der Stadt Göttingen", Göttingen 1962

图9（中）：G. Schwalbach. Source: GOETTINGA, urbs munitissima et splendida ..., an urban plan by Matthias Seutter, Augsburg (copper engraving from around 1750). Graphic design: Atelier K.-H. Fehrecke, commissioned by city of Göttingen construction authority, first published in 1971

图9（右）：Graphic: G. Schwalbach. Source: plan by municipal planning office of Göttingen 1893. Graphic design: Atelier K.-H. Fehrecke, commissioned by city of Göttingen construction authority. First published in 1971

图10, 48：Pesch & Partner: Architekten+Stadtplaner, Herdecke/Stuttgart: Gestaltungsfibel und -satzung Innenstadt Coesfeld (design manual and design statutes for inner-city Coesfeld), Herdecke 2006

图19, 21：© Geobasisdaten (spatial base data) Bundesland North Rhine-Westphalia, Bonn

图20:	© City of Dortmund, Vermessungs- und Katasteramt, license no.: 0901250
图22:	© ESRI Geoinformatik GmbH
图23:	colibri X7 protect, © Mettenmeier GmbH, D-Paderborn
图24:	Area use plan for the city of Drolshagen. 11th amended version, Drolshagen 2009
图29, 31, 33, 35, 39, 40, 42:	Pesch & Partner. Architekten+Stadtplaner, Herdecke/ Stuttgart: preliminary investigation plan and urban framework development plan for Remscheid-Rosenhügel, Herdecke 2005
图30:	Map showing overview of community of Remscheid, approx. 1868, Deutsches Werkzeugmuseum/Remscheid
图37:	Pesch & Partner. Architekten+Stadtplaner, Herdecke/ Stuttgart. Preliminary planning of Horster Strasse in Gelsenkirchen, Herdecke 2002
图41, 49, 54:	Schwalbach, G./University of Siegen: an investigation into the implications of demographic change for Drolshagen and its surrounding villages, Siegen 2009, Map used as a basis provided by the Katasteramt of the Olpe district.
图44, 56:	Pesch & Partner. Architekten+Stadtplaner, Herdecke/ Stuttgart: urban framework development plan for Mittlerer Bruchweg in Recklinghausen, Herdecke 2003
图45:	Rüdiger Brosk, Essen
图46:	Pesch & Partner. Architekten+Stadtplaner, Herdecke/ Stuttgart: Leitbild 2020: Göttingen stellt sich der Zukunft, Göttingen/Herdecke 2007 (city of Göttingen, FD Bodenordnung, Vermessung u. Geoinformation AZ: 622371)
图47, 52, 57, 59:	Pesch & Partner. Architekten+Stadtplaner, Herdecke/ Stuttgart: Leitbild 2020: Göttingen stellt sich der Zukunft, Göttingen/Herdecke 2007

其他图表和插图由作者本人提供。

作者简介

格里特·施瓦尔巴赫，建筑学学士，建筑师，德国锡根大学（University of Siegen）建筑与城市规划系城市规划和规划史方向助理研究员。